"*In* The Genie in Your Genes, *Dawson Church reveals, in a most powerful, practical, and hopeful way, the extraordinary link among body, mind, and spirit, and the undeniable impact of our thoughts and emotions on our everyday experiences. Anyone who reads this book and takes its remarkable information to heart will be equipped to draw to themselves a level and degree of health, abundance, and joy that far surpasses anything they could have imagined.*"

—Susan Schachterle, Author of *The Bitch, The Crone, and The Harlot*

"*Take yourself on a quantum leap into cutting edge science and discover how it will transform medicine and health care in the next decade. Dawson Church has mined the most revolutionary findings from diverse fields of science and practice and brilliantly synthesized them into a clearly-written manifesto for the enlightened care of the human body, mind, and community.*"

—Donna Eden & David Feinstein, Ph.D., Coauthors of the best-selling *Energy Medicine*

"*Once in a great while, an expansive book traces the connections between seemingly disconnected fields of science to produce a brilliant new synthesis. This work, linking genetics, electromagnetism, medicine and social change, is a monumental feat of scholarship and imagination, and provides a fascinating glimpse of the exciting possibilities that lie ahead. It points the way to a radical leap in our understanding of healing—and indeed the nature of human beings.*"

—Gary Craig, Originator of EFT, Coauthor of *The Promise of Energy Psychology*

"*The Genie in Your Genes is an astonishing look at how our how our beliefs, thoughts and emotions impact every aspect of our very being. If you want to know where medicine and psychology are headed in the next hundred years, and how to create extraordinary well-being in your life—right now—this is the book to read.*"

—Ray Dodd, Author of *Belief Works* and *The Power of Belief*

The Genie in Your Genes

Epigenetic Medicine
and the New Biology of Intention

Dawson Church, Ph.D.

www.EpigeneticMedicine.org

Elite Books
Santa Rosa, CA 95403
www.EliteBooksOnline.com

Cataloging-in-Publication Data

Church, Dawson, 1956-
The genie in your genes: epigenetic medicine and the new biology of intention /
Dawson Church.

p. cm.

ISBN-13: 978-1-60070-022-4
ISBN-10: 1-60070-022-5

1. Energy psychology. 2. Genetic regulation. 3. Epigenesis. 4. Healing—Psychological
aspects. I. Title.

RC489.E53C48 2007

616.89--dc22

2007009305

Copyright © 2007, Dawson Church

The information in this book is not to be used treat or diagnose any particular disease or any particular patient. Neither the authors nor the publisher is engaged in rendering professional advice or services to the individual reader. The ideas, procedures and suggestions in this book are not intended as a substitute for consultation with a professional health care provider. Neither the authors nor the publisher shall be liable or responsible for any loss or damage arising from any information or suggestion in this book.

All illustrations and photographs in this book are © istockphoto, © Dawson Church, or are copyright-free, except where noted in the endnotes.

Figures and quotes identified in the endnotes as drawn from *Energy Medicine in Therapeutics and Human Performance* are reprinted with permission of Elsevier, and are copyright © 2004 James Oschman.

Selections from *The Promise of Energy Psychology* identified in the endnotes, and the Appendix "The Basic Recipe on a Page," are used with permission of Tarcher/Penguin, and are copyright © 2005 David Feinstein, Gary Craig, and Donna Eden.

Excerpts from *The Psychobiology of Gene Expression* identified in the endnotes are used with permission of Norton, and are copyright © 2002 Ernest Rossi.

The Appendix "How To Choose a Practitioner" is reprinted from *Soul Medicine* ©2006 by C. Norman Shealy and Dawson Church and is used with permission. The chapter "Anatomy of a Monster" appeared in a different form in the anthology *The Heart of Healing,* ©2004 Elite Books.

The Appendix "Five Minute Energy Routine" is reprinted from *Energy Medicine Interactive* ©2006 by Donna Eden and David Feinstein and is used with permission.

Typeset in Hoefler Text and Capitals
Printed in USA
First Edition
Illustrations by Karin Kinsey
Cover Design by Robert Mueller
10 9 8 7 6 5 4 3 2 1

This book is dedicated to C. Norman Shealy, M.D., Ph.D., founder of the American Holistic Medical Association, cofounder of Holos University, and inspirer of several generations of nurses, physicians, researchers, and healing practitioners.

Contents

Acknowledgments

Many times a day I realize how much my own outer and inner life is built upon the labors of my fellow men, both living and dead, and how earnestly I must exert myself in order to give in return as much as I have received.

—Albert Einstein

I would like to express my deep appreciation to several brilliant authors, researchers, and doctors who provided the insights that made this present book possible.

First, to Norman Shealy, M.D., Ph.D., I am indebted for an education in *the critical role that electricity and magnetism play in all healing.* This is the first of several insights that led to the writing of this book. Norm is an expert in the subject; I coauthored a book with him called *Soul Medicine,* which stimulated my curiosity about the links between faith healing and electromagnetic conduction. Norm also chaired my dissertation committee, and has contributed mightily to the professional development of alternative medicine in many ways: through his insistence on rigorous documentation of "miraculous" cures; through his establishment of Holos University; through his founding of the American Holistic Medical Association (AHMA) and co-founding of the American Board for Scientific Medical Intuition (ABSMI); through the development of the SheliTENS treatment machine; and through his new book, *Life Beyond 100.* Norm has also mentored me, encouraged me, listened patiently (well, at least stoically) to me during my explication of ideas in process, and even—surely the supreme test of human courage—suffered through being a passenger in my car on the freeways of Los Angeles, California.

I would also like to thank Ernest Rossi, Ph.D., for his amazing book *The Psychobiology of Gene Expression.* Until I read his book, I did not realize *just how quickly genes may be activated by memories, feelings, thoughts,* and other aspects of the internal environment—some genes in *under two seconds.* This was the second great insight that motivated me to complete this book. Rossi applies these findings to psychotherapy in general and clinical hypnosis in particular, in ways that explain many mysteries of the therapeutic process. In the book above, Rossi has laid a foundation for a whole generation of researchers to investigate the effects on genes of mental and emotional shifts in the individual. Rossi's book gave me many research pointers and great inspiration.

Another book that contains stunning implications for medicine and research is *Energy Medicine in Therapeutics and Performance*. Written by James Oschman, Ph.D., this book synthesizes much other research in the field in a way that presents a compelling picture of some of the mechanisms by which energy creates healing. To Jim I am especially indebted for the third insight: that *the body's connective tissue system is a giant liquid crystal semiconductor*. Oschman's comprehensive survey of research in the field makes further research much easier. He has also done duty as a friend, cheerleader, and critical thinker in the development of this book, reviewing and correcting the chapters on connective tissue and its role in the transmission of information in the human body.

I would also like to thank those who have instructed me in Energy Psychology. This group of therapies are also sometimes known as Meridian Based Therapies, or MBTs. To these wise instructors I owe the discovery that *emotional trauma—even deep, ingrained, long-standing trauma—can be alleviated very quickly, often in just a few minutes*. I had great trouble believing this when I first heard about it, and my logical belief system still has trouble wrapping itself around the idea that a neurosis or psychopathology for which a patient might have spent an unsuccessful decade in psychotherapy can be permanently resolved in mere moments through Energy Psychology. Yet there are thousands of case histories, and a growing number of clinical studies, that show that this is often possible.

Therefore, I believe that Energy Psychology has a potential for the alleviation of human suffering that rivals any advance in psychology or medicine in the last five centuries. That's a huge claim, but it's not hyperbole. Try it for yourself; you'll be a convert within an hour—and it will change your life thereafter! Among the developers and popularizers of Energy Psychology are:

Gary Craig, the developer of Emotional Freedom Techniques (EFT), from whom many of the case histories in this book are drawn. Gary is a passionate advocate of these techniques, and has done more than any other single individual to advance the field. Over *half a million* people have now downloaded Gary's free EFT manual at www.emofree.com. That result (similar in quantity to the downloads of a new beta version of Microsoft Windows) is not the result of a slick marketing campaign—or any marketing campaign at all. It's all word of mouth: hundreds of thousands of mouths testifying to the remarkable power of EFT. Gary has also recorded an invaluable teaching series of DVDs, and they are an excellent introduction to the astounding potential of Energy Psychology, as well as a primer on how to start using it yourself. Thanks also to Gary for his wise personal counsel, his insights into how to use Energy Psychology to address even serious disease, and for his passionate advocacy of these methods, which have helped tens of thousands of people escape from long-standing patterns of distress.

I deeply appreciate David Feinstein, Ph.D., and Donna Eden, who together with Gary wrote a clear and compelling guide to this exciting new technique called *The Promise of Energy Psychology.* Thanks, too, to David Feinstein for suggesting the title of this book at a time when I was having trouble seeing the forest for the trees. I also appreciate the inspiration I derived from his seminal papers defining the field (see www.HandoutBank.org), and for his valuable feedback and friendship. David has also done a great deal to encourage professional standards of competence in the field, and to collect the science behind these methods into a coherent whole. David and Donna's book *Energy Medicine* is a best-seller, and one of the foundational texts for this field.

Daniel Benor, M.D., is a pioneering psychiatrist whose Wholistic Hybrid of EMDR and EFT (WHEE) technique is one of the simplest Energy Psychology techniques for daily personal and clinical practice.

His four volumes of *Healing Research,* by assembling so many studies in a single place, make a compelling case for the scientific basis of this work. Among the authors inspired by his work are Larry Dossey, M.D., whose bestselling books have done a great deal to open the minds of Western medical practitioners to the possibilities of alternative medicine. Among his inspirational activities, Dan chairs the Council on Healing, a cross-disciplinary body connecting practitioners from many fields, and cross-fertilizing their best ideas. By developing an MBT that is so fast and inconspicuous that even gang members in juvenile hall can use it, Dan has contributed an immensely practical dimension to Energy Psychology.

Fred Gallo, Ph.D., who coined the term *Energy Psychology* and whose book of the same name is a sound introduction for professionals, has developed tools and techniques that have benefited many. Fred's books are the foundational texts for the field, and by systematizing these treatments, then organizing them into a simple treatment system, he has given it a solid underpinning.

I also gratefully acknowledge Tapas Fleming, L.Ac., developer of the Tapas Acupressure Technique or TAT, among the easiest Energy Psychology techniques to learn, yet one which harnesses similar power to shift traumas. Her quiet wisdom, gentle compassion, and grounded teaching techniques have demystified Energy Psychology and healed many lives.

Though I have not yet met him, I acknowledge an immense debt of gratitude to Roger Callahan, Ph.D., the originator of Thought Field Therapy or TFT, and the author of *Tapping the Healer Within* and other books. He articulated and popularized his methods in the 1970s and 1980s, and development of these methods led to EFT and most other subsequent Energy Psychology techniques.

I am immensely grateful to David Gruder, Ph.D., cofounder of the Association for Comprehensive Energy Psychology (ACEP). By creating and leading an umbrella organization of pioneers in this field, he provided an environment ripe for the cross-fertilization of theories and practices, and continues to inspire audiences with the power of his insights. Larry Stoler, Ph.D., president of ACEP as of this writing, and Mary Sise, LCSW, Dorothea Hover-Kramer, Ed.D., and Gloria Arenson, MFT, past presidents, have all contributed to the vibrancy of the organization. Every conversation I have with Larry yields a fresh insight, and his wisdom, honesty, and enthusiasm have brought new energy to the organization. Under Larry's leadership, ACEP has instituted a Certification Program in Energy Psychology, identifying a uniform standard of knowledge and clinical practice, and training the first batch of graduates in this discipline.

I am also grateful to David for organizing the many steps in spiritual and psychological and emotional healing into a unified whole, and explaining to me, and others, how they fit into a grand developmental scheme. David's work is just beginning; I have predicted that the power of his insights will result in him one day being regarded as, "The Sigmund Freud of the twenty-first century"; we will see if this bold prediction holds true. David also authored many of the key documents in the development of the Energy Psychology field, including ACEP's professional code of ethics, which is a model of its kind.

This field owes a great debt to Francine Shapiro, Ph.D., who popularized Eye Movement Desensitization and Reprogramming (EMDR), and furthered this technique by insistence on rigorous clinical studies. I would also like to acknowledge all those who have served on ACEP's research committee, especially Greg Nicosia, Ph.D., who has midwifed several clinical studies of EFT and is playing a pivotal role in setting rigorous scientific standards for experiments.

Michael Mayer, Ph.D., a therapist in private practice in the San Francisco Bay Area, and Qigong master, first made me aware that what we now call Energy Psychology—and are tempted to think of as a new therapeutic approach—is in fact simply a *new application of the ancient principles of Qigong*, which are several thousand years old.

My great appreciation also goes to my wonderful friends Bruce Lipton, Ph.D., and his partner Margaret Horton. Bruce's book *The Biology of Belief,* which I published, hit various best-seller lists in 2005 when it came out, has sold over 100,000 copies, and catalyzed the idea in popular awareness that *genes are merely blueprints, and that it's the environment outside our cells that determines genetic expression.* Bruce has also advanced the idea that the cell's "brain" is its membrane, for it is the membrane, influenced again by the environment, that admits the signals that trigger the processes of life. Bruce's book created a fertile public awareness for these radical new ideas, in which books like this one, and *Soul Medicine,* can bear fruit.

I deeply appreciate the editing of John Travis, M.D., coauthor of *The Wellness Workbook,* who gave me the benefit of his brilliant and perceptive mind. He went through this manuscript with his eagle eye prior to sending it out for peer review, and corrected some of the key concepts, as well as fine-tuning the language of the text. He challenged my hidden assumptions, as well as picking apart any imprecise science and reassembling it correctly. The final draft has benefited from his decades of experience and compendious knowledge of both conventional and alternative medicine. His work in bringing awareness to childrearing through the Alliance for Transforming the Lives of Children (www.aTLC.org), which envisions a world where "every child is wanted, welcomed, loved, and valued," and where "every family is prepared for, and supported in, practicing the art and science of nurturing children," brings awareness of the many concrete steps we can to create a life-affirming culture of childraising.

I have also received encouragement from some my heroes: Larry Dossey, M.D., very kindly granted me interviews, or permission to publish his work, in several of my anthologies. Caroline Myss, Ph.D., did the same, and also encouraged and furthered the publication of *Soul Medicine.* Barbara Marx Hubbard, one of the shining lights of our generation, first traced for me the implications of some of these ideas for human and planetary evolution. Bob Nunley, Ph.D., and Ann Nunley, Ph.D., besides being instrumental, along with Norm Shealy, in founding Holos University, also developed a technique that

I have found most powerful, and which incorporates the most powerful transformative techniques of transpersonal psychology, called Inner Counselor.

I am also extremely grateful to the salespeople who believed in this book and represented effectively to bookstores; with over 100,000 new titles published each year, it's hard to stand above the crowd unless you have passionate advocates in the sales cycle. I particularly appreciate Eric Kampmann, my friend and supporter, president of Midpoint Trade Distribution; also Gail Kump, the deft and articulate cheerleader who persuaded Barnes and Noble to put the book on the front table of all their stores the week it was released, Chris Bell, who encouraged Borders to take a serious look at the book, Margaret Queen, whose kindness, and steady support for all my books with book wholesalers, has been a pillar of strength, and John Teall, whose cheerful competence with Amazon and other online accounts has made dealing with them a pleasure.

Books and articles that concern themselves with health and healing go through the process of "peer review," in which colleagues and experts read the draft and comment on the material from the perspective of their expertise. I deeply appreciate those who read all or part of this manuscript and offered their suggestions and advice. When an author sends rough drafts into the world, he feels like an artist unveiling a sculpture. The eyes of the audience are on the sculpture. But the eyes of the author are on the audience! Until you gauge the reaction of your audience, you don't know whether you're unwrapping a masterpiece or a monstrosity. The feedback of colleagues and friends is an essential part of spotting the weak links in an argument and polishing it up. I deeply appreciate the feedback of these generous souls, including:

Scott Anderson, M.D.
Daniel Benor, M.D.
Pat Butler, M.A.
Gary Craig
David Feinstein, Ph.D.

Fred Gallo, Ph.D.

David Gruder, Ph.D.

Jeanne House, M.A.

Gail Ironson, M.D., Ph.D.

Ken Jenkins, Ph.D.

Mark P. Line

Bruce Lipton, Ph.D.

Greg Nicosia, Ph.D.

James Oschman, Ph.D.

Kent Peterson, M.D.

Tamala Pulling

Leane Roffey, Ph.D.

Beverly Rubik, Ph.D.

Brenda Sanders, Ph.D.

Norman Shealy, M.D., Ph.D.

Larry Stoler, Ph.D.

John Travis, M.D.

Bernard Williams, Ph.D.

They brought both healthy skepticism and welcome encouragement to the project. They pointed out newer studies I was unaware of, holes in my early reasoning, areas where the arguments needed strengthening, gaps in my logic, and other defects that required correction before publication. Collectively they represent a database of knowledge and experience many orders of magnitude greater than mine, and I am humbled by their generosity in donating their time and expertise to critiquing this work.

It is impossible for me to find the quiet creative space to write books or articles in the pressured and deadline-driven publishing office where I work nine to five (or more usually five to nine—five A.M. to nine P.M.); getting away on a cruise, a "vacation," or to a spa has been the only thing that made it possible. So besides writing part of this book while looking out the window of the library on the cruise ship *Dawn Princess,* I wrote parts at Harbin Hot Springs, parts at Tahoe State Park, and parts at the Prince Kuhio Resort in Kauai,

Hawaii. Other people took good care of me at those times, enabling me to do my work, or took care of the office so I could leave: my conspirators at Elite Books, Courtney Arnold and Jeanne House, my daughter Angela, the staff at Harbin, and the staff of Princess Cruise Lines. This book has been the biggest single project of my life, and I appreciate all who made it possible.

—Dawson Church

1

Epigenetic Healing

Imagination is everything. It is the key to coming attractions.

—Albert Einstein

One morning a part-time employee walked into my office, at the start of a busy workday. Her name was Anabelle. Tall, slender, well dressed and poised, with commanding blue eyes and a sharp intelligence, Anabelle was a force to be reckoned with. Except for that morning...

The moment I opened the office door, Anabelle appeared to be in such obvious distress that it was clear that she was falling to pieces, emotionally. Instead of going to a desk, I took her elbow and guided her to a chair in the waiting room, where, without much prompting, she poured out her sad story.

Her stepfather, Jack, had abused her, verbally and physically, from the age of around seven. She ran away from home at the age of fifteen and rarely visited her parents after that. The last time Jack had hit her was on one of those visits. At that time, she was twenty-one years old. She never went back to the house; it had now been over fifteen years since the last incident.

In the previous day's mail, she had received an embossed formal invitation to a family reunion. It had been sent to her by her mother, who was still married to Jack, and who was arranging the get-together. Anabelle knew that Jack would be there.

"How could my mother send me this invitation, knowing I'd have to see him again?" she wailed. That slip of paper summarized the whole nightmare of abuse for Anabelle—plus what she saw as her mother's condoning Jack's behavior.

I asked her if I could perform an energy intervention with her that might make her feel better. She nodded wordlessly. I then asked her to remember the moment she had opened the envelope and read the invitation, and feel where in her body the sensation of distress was most concentrated.

She responded, "My tummy" and pointed at her solar plexus. I then asked her to rate her distress, as she thought again about the scene, on a scale of one to ten, with one being calm and ten being as upset as she could possibly be. "I'm a *ten!*" she said through tight lips, with flushed cheeks, her voice rising emphatically.

I performed a very fast and basic emotional energy release technique, used by thousands of doctors and therapists worldwide, and now visible on the radar of cutting-edge researchers. The entire process took less than two minutes. "Now think back on the moment you opened the envelope," I asked her. "Remember the scene. Feel your tummy. Then tell me how upset you feel, on a scale of one to ten."

"Zero." She shrugged, and looked at me with calm puzzlement. Then she added, "It was just a scrap of paper, after all."

I sat opposite Anabelle, stunned and mute. My logical mind—never usually at a loss for thoughts, judgments, observations, or words—grasped vainly for rational reasons that might explain this astonishing change. Like a fish out of water, my mouth worked and my eyes widened as I shuffled desperately through my mental cue cards for rational explanations. Even after witnessing many of these interventions, and doing them myself often, my left brain still has trouble grasping the evidence before my eyes.

Turning Gene Research into Therapy

From all around the world, in virtually every field of the healing arts—from psychiatrists to doctors to psychotherapists to sports physiologists to social workers—stories like this are being told, as the world of psychology and medicine begins to awaken to the potential of energy medicine and its effects on the expression of our DNA. They are the first loud reports of a revolution in treatment destined to change our entire civilization, reaching into every corner of medicine and psychology...and beyond them into the structures of society itself. In the space of one generation we have discovered, or rediscovered, techniques that can make us happier, less stressed, and much more physically healthy—safely, quickly, and without side effects. Techniques from energy medicine and Energy Psychology can alleviate chronic diseases, shift autoimmune conditions, and eliminate psychological traumas with an efficiency and speed that conventional treatments can scarcely touch.

The implications of these techniques—for human happiness, for social conflicts, and for political change—promise a radical positive disruption in the human condition, one that goes far beyond health care. They have the promise of affecting society as profoundly as the rediscovery of mathematical and experimental principles during the Renaissance changed the course of European civilization. And they are at the cutting edge of science, as experimental evidence stacks up to provide objective demonstration of their effectiveness.

Along with the practical evidence accumulating from pioneers in this new medicine and new psychology, scientists are discovering the precise pathways by which changes in human consciousness produce changes in human bodies. As we think our thoughts and feel our feelings, our bodies respond with a complex array of shifts. Each thought or feeling unleashes a particular cascade of biochemicals in our organs. Each experience triggers genetic changes in our cells.

The Dance of Genes and Neurons

These new discoveries have revolutionary implications for health and healing. Psychologist Ernest Rossi begins his authoritative text *The Psychobiology of Gene Expression* with a challenge: "Are these to remain abstract facts safely sequestered in academic textbooks, or can we take these facts into the mainstream of human affairs?"[1]

The Genie in Your Genes takes up Rossi's challenge, believing that it is essential that this exciting genetic research progress beyond laboratories and scientific conferences and find practical applications in a world in which many people suffer unnecessarily. Rossi explores "how our subjective states of mind, consciously motivated behavior, and our perception of free will can modulate gene expression to optimize health."[2] Nobel prizewinner Eric Kandel, M.D., believes that in future treatments, "Social influences will be biologically incorporated in the altered expressions of specific genes in specific nerve cells of specific areas of the brain."[3] Brain researchers Kemperman and Gage envision a future in which the regeneration of damaged neural networks is a cornerstone of medical treatment, and doctors' prescriptions include, "modulations of environmental or cognitive stimuli" and "alterations of physical activity." In other words, when the doctor of the future tears a page off her prescription pad and hands it to a patient, the prescription might well be—instead of, or in addition to, a drug—a particular therapeutic belief or thought, a positive feeling, a gene-enhancing physical exercise, an act of altruism, or an affirmative social activity. Research is revealing that these activities, thoughts, and feelings have profound healing and regenerative effects on our bodies, and we're now figuring out how to use them therapeutically.

The Dogma of Genetic Determinism

This picture of a genetic makeup that fluctuates by the hour and minute is at odds with the picture engrained in the public mind: that genes determine everything from our physical characteristics to our behavior. Even many scientists still speak from the assumption that our genes form an immutable blueprint that our cells must forever follow. In her book *The Private Life of the Brain,* British research scientist and Oxford don Susan Greenfield says, "the reductionist genetic train of thought fuels the currently highly fashionable concept of a gene for this or that."[5]

Niles Eldredge, in his book *Why We Do It,* says, "genes have been the dominant metaphor underlying explanations of all manner of human behavior, from the most basic and animalistic, like sex, up to and including such esoterica as the practice of religion, the enjoyment of music, and the codification of laws and moral strictures.... The media are besotted with genes...genes have for over half a century easily eclipsed the outside natural world as the primary driving force of evolution in the minds of many evolutionary biologists."[6]

Medical schools have had the doctrine of genetic determinism embedded in their teaching for decades. The newsletter for the students at the Health Science campus of the University of Southern California proclaims, "Research has shown that 1 in 40 Ashkenazi women has defects in two genes that cause familial breast/ovarian cancer...."[7] Unexamined beliefs in this or that gene causing this or that condition are part of a the foundation of many scientific disciplines in our society.

Such assumptions can be found in stories like one that aired on National Public Radio on October 28th, 2005. The announcer declared: "Scientists today announced they have found a gene for dyslexia. It's a gene on chromosome six called DCDC2." The *New York Times* ran a similar story the following day, under the headline, "Findings Support That [Dyslexia] Disorder Is Genetic." Other media picked up the story, and the legend of the primacy of DNA was reinforced.

There's only one problem with the legend: it's not true.

Actually, there's a second major problem with the legend: it locates the ultimate power over our health and wellbeing in the untouchable realm of molecular structure, rather than in our own consciousness. In her book *The DNA Mystique,* Dorothy Nelkin states that, "In a diverse array of popular sources, the gene has become a supergene, an almost supernatural entity that has the power to define identity, determine human affairs, dictate human relationships, and explain social problems. In this construct, human beings in all their complexity are seen as products of a molecular text...the secular equivalent of a soul—the immortal site of the true self and deter-miner of fate."[8]

In reality, genes contribute to our characteristics but do not determine them. Blair Justice, Ph.D., in his book *Who Gets Sick,* observes that, "genes account for about 35% of longevity, while lifestyles, diet, and other environmental factors, including support systems, are the major reasons people live longer."[9] The percentage by which genetic predisposition affects various conditions varies, but it is rarely 100%. The tools of our consciousness—including our beliefs, prayers, thoughts, intentions, and faith—often correlate much more strongly with our health, longevity, and happiness than our genes do. Larry Dossey, M.D., observes, "Several studies show that what one *thinks* about one's health is one of the most accurate predictors of longevity ever discovered."[10] Studies show that a committed spiritual practice and faith can add many years to our lives, regardless of our genetic mix.[11]

How did the dogma that DNA holds the blueprint for develop-ment become so firmly enshrined? In his entertaining book *Born That Way,* medical researcher William Wright gives a detailed history of the rise to supremacy of the idea that genes contain the codes that control life—that we are who we are, and we do what we do, because we were simply "born that way."[12] We often hear phrases like "She's a natural born athlete," or "He's a born loser," or "She has good genes," to explain some aspect of a person's behavior. The idea of genetic disposition has moved far beyond the laboratory to become deeply entrenched in our popular culture.

Lee Dugatkin, professor of biology at the University of Louisville, points out that after the basic rules governing the inheritance of char-acteristics across generations were made by Mendel, and the structure of the DNA molecule was discovered, scientists became convinced that the gene was the "means by which traits could be transmitted across generations. We see this trend continuing today in research labs throughout the world as well as in the media in reports of genes for schizophrenia, genes for homosexuality, genes for alcoholism, and so on. Genes for this, genes for that."[13] Researcher Carl Ratner, Ph.D., of Humboldt State University draws the following analogy: "Genes may directly determine simple physical characteristics such as eye color. However, they do not directly determine psychological phenomena. In the latter case, genes produce a potentiating substratum rather than particular phenomena. The substratum is like a Petri dish which forms a conducive environment in which bacteria can grow, however, it does not produce bacteria."[14]

Yet, since the 1970s, researchers have been turning up findings that are at odds with the prevailing mindset. They have accumulated an increasing number of findings that behaviors aren't just transmitted genetically across generations; they may be newly developed by many individuals during a single generation. While the process of genetic evolution can take thousands of years, as genes throw off mutations that are sometimes successful, and often not, evolution through experience and imitation can occur within minutes—and *then* be passed on to the next generation.

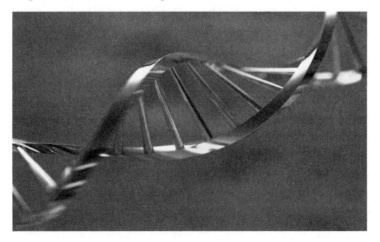

The DNA spiral has become a defining icon of our civilization

Edward O. Wilson, the father of sociobiology, hinted at the very end of the twenty-fifth anniversary edition of his tremendously influential book, *Sociobiology*, that, in future research, "Learning and creativeness will be defined as the alteration of specific portions of the cognitive machinery regulated by input from the emotive centers. Having cannibalized psychology, the new neurobiology will yield an enduring set of first principles for sociology.... We are compelled to drive toward total knowledge, right down to the levels of the neuron and gene."[15] The notion that the genes in the neurons of our brain can be activated by input from our emotive centers is a big new idea, and indicates a degree of interconnection and feedback at odds with the straight-line, cause-and-effect model of genetic causation.

As well as beings of matter, we are beings of energy. Electromagnetism pervades biology, and there is an electromagnetic component to every biological process. While biology has been largely content with chemical explanations of how and why cells work, there are many tantalizing preliminary research findings that show that electromagnetic shifts accompany virtually every biological process. The energy flows in neurons and genes interact with their every process. "There is matter and energy galore flowing through biological systems," says Eldredge. "But it is in the bodies of organisms and their interactions with other organisms and the physical world, in the context of ecosystems, where all that matter and energy flows. Genes, in contrast, are about storage and utilization of information." Researching genes without looking at the energy component of DNA is like studying a computer hard drive without plugging in the power cable. Hard drives are composed of thousands of sectors, substructures that store information.[16] You can develop impressive theories about why the storage device is constructed the way it is, and the interesting way in which the sectors are arranged, but until you plug the hard drive in and watch it functioning in the context of the energy flow that animates it, you have a very incomplete picture of the way it works.

Death of a Dogma

The idea that genes are the repositories of our characteristics is also known as the Central Dogma. The Central Dogma was propounded by one of the discoverers of the helicular structure of DNA, Sir Francis Crick. He first used the term in a 1953 speech, and restated it in a paper in the journal *Nature,* entitled, Central Dogma of Molecular Biology.[17] Yet for some thirty years, scientists have been turning up anomalous data that is not compatible with the Central Dogma. The outcomes of these experiments require much more complex interactions than genetic determinism can explain.[18]

One of many problems with the dogma, for instance, is that the number of genes in the human chromosome is insufficient to carry all the information required to create and run a human body. It isn't even a big enough number to code for the structure (let alone function) of one complex organ like the brain. It also is too small a number to account for the huge quantity of neural connections in our bodies.[19] Two eminent professors express it this way: "Remembering that the information in the human genome has to cover the development of all other bodily structures as well as the brain, this is not a fraction of the information required to structure in detail any significant brain modules, let alone for the structuring of the brain as a whole."[20]

The Human Genome Project initially was focused on cataloging all the genes of the human body. At the beginning of the 1990s, the original researchers expected to find at least 120,000 genes, because that's the minimum they projected it would take to code all the characteristics of an organism as complex as a human being. Our bodies manufacture about 100,000 proteins, the building blocks of cells. All of those 100,000 building blocks must be assembled with precise coordination in order to support life. The working hypothesis at the start of the Human Genome Project was that there would be a gene that provided the blueprint to manufacture each of those 100,000 proteins, plus another 20,000 or so *regulatory genes* whose function was to *orchestrate* the complex dance of protein assembly.

The further the project progressed, the smaller the estimates of the number of genes became. When the project finished its catalog, they had mapped the human genome as consisting of just 23,688 genes. The huge symphony orchestra of genes they had expected to find had shrunk to the size of a string quartet. The questions that this small number of genes gives rise to are these: If all the information required to construct and maintain a human being—or even one big instrument, such as the brain—is not contained in the genes, where does it come from? And who is conducting the whole complex dance of assembly of multiple organ systems? The focus of research has thus shifted from cataloging the genes themselves to figuring out how they work in the context of an organism that is in "a state of *systemic cooperation* [where] every part knows what every other part is doing; every atom, molecule, cell, and tissue is able to participate in an intended action."[21]

The lack of enough information in the genes to construct and manage a body is just one of the weaknesses of the Central Dogma. Another is that genes can be activated and deactivated by the environment inside the body and outside of it. Scientists are learning more about the process that turns genes on and off, and what factors influence their activation. We may have lots of information on our hard drives, but at a given time we will be utilizing only part of it. And we may be changing the data as well, like revising a letter before we send it to a friend. One of the factors that affect which genes are active is our experience, a fact completely incompatible with the doctrine of genetic determinism.

Yet our experiences themselves are just part of the picture. We take facts and experiences and then assign meaning to them. What meaning we assign, mentally, emotionally, and spiritually, is often as important to genetic activation as the facts themselves. We are discovering that our genes dance with our awareness. Thoughts and feelings turn sets of genes on and off in complex relationships. Science is discovering that while we may have a fixed set of genes in our chromosomes, which of those genes is active has a great deal to do with our subjective experiences, and how we process them.

Our emotions and behavior shape our brains as they stimulate the formation of neural pathways that either reinforce old patterns or initiate new ones. Like widening a road as traffic increases, when we think an increased flow of a thoughts on a topic, or practice an increased quantity of an action, the number of neurons our bodies requires to route the information increases. In just the way our muscles bulk up with increased exercise, the size of our neural bundles increases when those pathways are increasingly used. So the thoughts we think, the *quality of our consciousness,* increases the flow of information along our neural pathways. According to Ernest Rossi, "we could say that *meaning* is continually modulated by the complex, dynamic field of messenger molecules that continually replay, reframe, and resynthesize neuronal networks in ever-changing patterns."[22] In the succinct words of another medical pioneer, "Beliefs become biology"—in our hormonal, neural, genetic, and electromagnetic systems, plus all the complex interactions between them.[23]

The Inner and Outer Environment

Memory, learning, stress, and healing are all affected by classes of genes that are turned on or off in temporal cycles that range from one second to many hours. The *environment* that activates genes includes both *the inner environment*—the emotional, biochemical, mental, energetic, and spiritual landscape of the individual—*and the outer environment.* The outer environment includes the social network and ecological systems in which the individual lives. Food, toxins, social rituals, and sexual cues are examples of outer environmental influences that affect gene expression. Researchers estimate that "approximately 90% of all genes are engaged...in cooperation with signals from the environment."[24]

Our genes are being affected every day by the environment of our thoughts and feelings, as surely as they are being affected by the environment of our families, homes, parks, markets, churches, and offices. Your system may be flooded with adrenaline because a mugger is running toward you with a knife. It may also be flooded with adrenaline because of a stressful change at work. And it may be flooded with adrenaline in the absence of any concrete stimulus other than the thoughts you're having about the week ahead—a week that hasn't happened yet, and may never happen. Let's take a look at the evolutionary purpose of these physiological events, and whether they're *adaptive* (helpful to your body) or *maladaptive* (harmful to your body).

Scenario One: Ten thousand years ago, when a mugger (or a member of a hostile tribe) ran at you with a sharp blade, you quickly took action. Your blood flowed away from your digestive tract toward your muscles. Your brain became hyperactive and your reproductive drive shut down. Thousands of biochemical changes took place in all the cells of your body within a couple of seconds, enabling you to run away from the attacker, or defend yourself.

You were already one of a select group of humans who had survived the dangers of a hostile world long enough to breed. Over tens of thousands of years, those with quick responses had survived long enough to produce offspring, and those with slow responses died before they could breed. So by the time a hostile tribesman ran at you with a blade, those thousands of years of evolutionary weeding had already produced a human admirably suited to fight or flee. The changes that occurred in your body in response to a threat were *adaptive;* they were *useful* adaptations for survival.

Scenario Two: Fast-forward ten thousand years. You're in a meeting that includes all the employees of your company. The firm has just been bought by a competitor. You know that the new owner is going to consolidate the work force. They aren't going to need everybody.

The manager of your division announces that after the meeting, when you return to your desk, you'll find either a pink slip, indicating that you're terminated, or nothing at all, meaning that you've survived the purge. Those who find pink slips are instructed to immediately clear out their desks and report to personnel for a severance check.

Suddenly, there are two tribes in the room: those that will survive, and those that will not. Worse, nobody but the manger knows who's in what tribe. The stress level in the room is unbearable. Who is your enemy? Who is your ally? You have no idea. You walk back to your desk, dreading what you will see, and dreading the lineup of fired and retained employees you will witness in the next hour.

Your desk has no pink slip. Neither does that of Harry, who works across from you. Suddenly you realize that the downsizing means you'll be thrown cheek to jowl with Harry, who though another survivor, is an incompetent liar. You look across to Helen's desk, and you see a pink slip. Helen is the most talented person in the building, someone on whom you've secretly depended for your success. Because her verbal skills are poor, the management failed to realize that she's indispensable. You realize your job has just become a lot worse, yet you will cling to it like a *Titanic* survivor gripping the last life jacket.

You've been working sixty-hour weeks for the last six months, suspecting that this Damocles' sword will eventually fall. Your body has been ready for fight or flight for all that time, not knowing what your fate will be. The employment market is tight; you know that many of the employees fired today will have to take Draconian pay cuts in menial new jobs.

Even before the current crisis, your body was in fight-or-flight mode as you climbed the corporate ladder. Today, it's on high alert. Your mouth is dry. You're so tense you could put your fist through the wall. You can't wait to get out of the office and have a few beers to unwind. Yet you know that tomorrow you'll be back at your desk—and now you'll have a huge new portion of the work that management has reassigned from the fired employees.

Scenario Three: It's Sunday evening. You've had a good weekend. You unwound by griping to your spouse on Friday night—then by playing baseball with your kids in the park on Saturday morning. You went to a movie with another couple on Saturday night, a comedy, and enjoyed a lot of laughs. You had sex with your partner after you got home. Church was good on Sunday morning, and you saw all your old friends there and had a chance to socialize.

You're sitting on the porch with a beer, and you suddenly realize you're going to have to go back into that hellhole of a job in just a few hours. Your stomach knots. Your jaw clenches. You crush the beer can. You start thinking of the injustices of the previous week, wondering how you escaped the axe. Didn't management see the glaring errors in your performance? You grind your teeth as you think of the injustice of them firing Helen, after she's kept the whole division going—in her quiet way—for years. What ingratitude! What blindness! What ineptitude! How did those morons get to be managers in the first place?

Can you escape? No chance, the money's good, the pension plan's good, and no other job has comparable medical benefits—vision and dental too, plus it covers the kids, for God's sake! Do you want to be pounding the pavement looking for a job like Helen will be doing tomorrow? God forbid!

Your Body Reads Your Mind

Scenario Two and Three are—in terms of what you're doing to your body—*maladaptive* responses. "Maladaptive" means that they aren't helping you; they're responses to stress that are hurtful to you. All the stress hormones are flowing, just as they were in Scenario One, but they're doing your body no practical good. No promotion will come as a result of you overloading your system with cortisol, one of the primary stress hormones. You won't feel better after being high on adrenaline and norepinephrine, two others.

What *will* happen, though, is that the circulation of these stress hormones through your system on a regular basis will compromise your immune system, weaken your organs, age you prematurely, and contribute to activating genes that worked perfectly well for the

36

caveman in Scenario One, but are counterproductive to the modern person in Scenarios Two and Three. Herbert Benson, M.D., president of Harvard Medical School's Mind-Body Medical Institute, says, "The stressful thoughts that lead to the secretion of stress-related norepinephrine impede our evolutionary-derived natural healing capacities. These thoughts are often only in our minds, not a reality."[25] According to another report, "Bruce McEwen, Ph.D., director of the neuroendocrinology lab at Rockefeller University in New York, says cortisol wears down the brain, leading to cell atrophy and memory loss. It also raises blood pressure and blood sugar, hardening arteries and leading to heart disease."[26]

So while the fight-or-flight response may have been adaptive ten thousand years ago, with Mother Nature cheering you on, today it's often maladaptive, and Mother Nature is saying, "Stop! You're ruining your body!" The trouble is that major evolutionary changes take a long time—sometimes thousands of years—and modern humans are having difficulty making adaptations in the short space of a single lifetime. We try and change our stress-addicted patterns in various ways. But the counteracting experiences we attempt—attending a four-evening stress clinic at the local hospital, a self-improvement workshop at a personal growth center, a weekend retreat at a church camp, or sitting in a Zen monastery for a few days—are like a tissue in a hurricane when compared to the evolutionary forces hardwired into our physiology.

Biochemically speaking, your body *cannot tell the difference* between the injection of chemicals that is triggered by an *objective* threat—the tribesman running at you with a spear—and a *subjective* threat—your resentment toward management. The biochemical and genetic effects, as far as your body is concerned, are the same. Your body can't tell that one experience is a physical reality, and the other is a replay of an abstract mental idea. *Both* are creating a chemical environment around your cells that is full of signals to your genes, several classes of which activate the proteins associated with healing.

A researcher observes: "Our body doesn't make a moral judgment about our feelings; it just responds accordingly."[27]

The understanding that much of our genetic activity is affected by factors outside the cell is a radical reversal of the dogma of genetic determinism, which held for half a century that who we are and what we do is governed by our genes. Research is showing a much more interconnected reality in which our *consciousness* plays a primary role.

Recent studies performed by Ronald Glaser, of the Ohio State University College of Medicine, and psychologist Janice Kiecolt-Glaser investigated the effect that stress associated with marital strife has on the healing of wounds, a significant marker of genetic activation. The researchers created small suction blisters on the skin of married test subjects, after which each couple was instructed to have a neutral discussion for half an hour. For the next three weeks, the researchers then monitored the production of three of the proteins that our bodies produce in association with wound healing. They then instructed the same couples to discuss a topic on which they disagreed. Research staff was present during both the neutral discussion and the disagreement.

The researchers found that the expression of these healing-linked proteins was depressed in those couple who had a fight. Even those couples who had a simple discussion of a disagreement, rather than a full-fledged verbal battle, showed slower healing of their wounds. But in couples who had severe disagreements, laced with put-downs, sarcasm, and criticism, wound healing was slowed by some 40%. They also produced smaller quantities of the three proteins. "'These are minor wounds and brief, restrained encounters. Real-life marital conflict probably has a worse impact,' Kiecolt-Glaser adds. 'Such stress before surgery matters greatly,' she says, and the effect could apply to healing from any injury. In earlier studies done by Kiecolt-Glaser, hostile couples were most likely to show signs of poorer immune function after their discussions in the lab. Over the next few months, they also developed more respiratory infections than supportive spouses."[28] Rossi says, "throughout the body's entire somatic network, emotions are triggering hormonal and genetic responses."[29] The genetic effects from such environmental experiences can, in some cases, make the difference between life and death. Pharmacologist Connie Grauds, R.Ph., in her book *The Energy Prescription,* says that, "An undisciplined mind leaks vital energy in a continuous stream of thoughts, worries, and skewed perceptions, many of which trigger disturbing emotions and degenerative chemical processes in the body."[30]

Over two thousand years ago, the Buddha declared: "We are formed and molded by our thoughts. Those whose minds are shaped by selfless thoughts give joy when they speak or act. Joy follows them like a shadow that never leaves them." Today's research is reinforcing what wise students of the human condition have known for millennia. Neuroscientist Candace Pert, Ph.D., tells us that, "the molecules of our emotions share intimate connections with, and are indeed inseparable from, our physiology.... Consciously, or more frequently, unconsciously, we choose how we feel at every single moment." Practices for health and wellbeing that were once the exclusive prescriptions of sages and priests are now being reinforced by geneticists and neurobiologists.

In the tales of the Arabian Nights, when Aladdin rubbed the magic lamp, the genie appeared and granted him three wishes. In the story, once he makes his wishes, the magic vanishes. He had to think long and hard on which three things he chose to wish for.

In the real world, given the lamp of our understanding and the genie in our genes, we have an unlimited supply of wishes. Whatever wishes we put into the lamp manifest genetically. If we fill our lamps with healing words, our genes rush to fulfill our wishes—within seconds. If, like the couples in the wound study above, we fill our lamps with poison, we damage the ability of our inbred genetic servants to heal us. While the mechanisms by which such differences occurred may have seemed like magic when viewed through the lens of allopathic medicine—the conventional system of treating symptoms with agents that produce an opposing effect—the results are not. In the coming pages, we will look, in detail, at the precise genetic and electromagnetic mechanisms that make such healing magic not only possible, but scientifically predictable.

Steps in Genetic Expression

The process by which a gene produces a result in the body is well mapped. Signals pass through the membrane of each cell and travel to the cell's nucleus. There, they enter the chromosome and activate a particular strand of DNA.

Around each strand of DNA is a protein "sleeve." This sleeve serves as a barrier between the information contained in the DNA strand and the rest of the intracellular environment. In order for the blueprint in the DNA to be "read," the sleeve must be unwrapped. Unless it is unwrapped, the DNA strand cannot be "read," or the information it contains acted upon. Until the information is unwrapped, the blueprint in the DNA lies dormant. That blueprint is required by the cell to construct other proteins that regulate virtually every aspect of life.

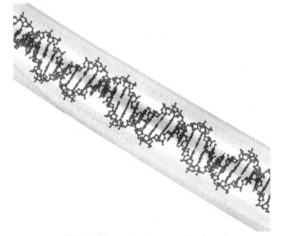

DNA blueprint in protein sheath

When a signal arrives, the protein sleeve around the DNA unwraps and, with the assistance of RNA, the DNA molecule then replicates an intermediate template molecule. The blueprint that has up to this point been concealed within the sleeve can now be acted upon. This is what scientists mean when they say that a gene *expresses.* The genetic information contained in the chromosome has gone from being a dormant blueprint into *active expression,* where it *creates other actions* within the cell by constructing, assembling, or altering products. The DNA blueprint that has up to this point been inert, concealed within the sleeve, is now revealed, providing the basis for cellular construction. Just as an architect's blueprint contains the information to build a building, the chromosomes contain the blueprints to construct aggregations of molecules. Until the architect's blueprint has been removed from its sheath, unrolled, laid flat on the builder's table, and used to guide construction, it is simply dormant potential. In the same way, the blueprints in our genes are dormant potential until the genes express and are used to guide the construction of the proteins that carry out the constructive tasks of life.

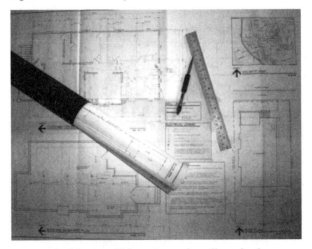

Architectural blueprint and cardboard tube

Proteins are the building blocks used by our bodies for every function they perform. Proteins control the responses of our immune systems, form the scaffolding that supports the structure of each cell, provide the enzymes that catalyze chemical reactions, and convey information between cells—among many other functions. If DNA is the blueprint, then RNA comprises the working drawings required for construction and proteins are the materials used in construction.

41

They are assembled into a coherent structure by the instructions in our DNA. That structure is not only our *anatomy*—the physical form of our bodies; it is also our *physiology*—the complex dance of cellular interactions that differentiate a live human being from a dead one. A corpse has anatomy, but no physiology. Proteins are used in every step of our physiology; the word "protein" itself is derived from the Greek word *protas,* meaning "of primary importance."

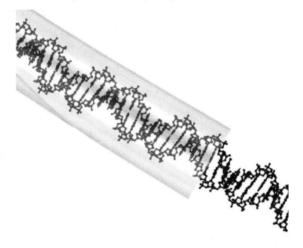

Protein sheath opens to permit gene expression

This whole chain of events starts with a *signal.* The signal is delivered through the cell membrane to the protein sleeve, which then unwraps in order to let the information in the gene move from potential (like an unbuilt building) to expression (like a finished sky-scraper). And while scientists have mapped each part of the process of gene expression and protein assembly, comparatively little atten-tion has been paid to the signals, the source of initiation for the whole process. Ignorance of the signal required to take the blueprint out of the tube is what has allowed several generations of biologists to assume that all you needed to start construction was the blueprint, giving rise to genetic determinism.

Signals From Outside the Cell

Stem cells are undifferentiated cells, "blanks" that the body can make into muscle, bone, skin, or any other type of cell. Like a piece of putty, they can be formed into whatever kind of cell the body needs. When you cut your hand and your body needs to repair the break in the skin, the trauma sends a signal to the genes associated with wound healing. These genes express, stimulating stem cells to turn themselves into healthy, fully functional skin cells. The signal results in the putty being formed into a useful shape. Such processes are occurring all over our bodies, all the time: "Healing via gene expression is documented in stem cells in the brain (including the cerebral cortex, hippocampus, and hypothalamus, muscle, skin, intestinal epithelium, bone marrow, liver, and heart."[31]

When there is interference with this signal, which in the wound healing studies comes from the emotional states of angry subjects, the stem cells don't get the message clearly. Not enough putty is turned into useful shapes, or the process of molding the putty takes a long time, because the body's energy is instead being gobbled up dealing with the angry emotion. Wound healing is compromised.

Notice that these *signals do not come from the DNA;* they come from *outside* the cell. The signals tell the proteins surrounding the DNA strands to unwrap and allow healing to begin. In the journal *Science,* researcher Elizabeth Pennisi writes, "Gene expression is not determined solely by the DNA code itself but by an assortment of proteins and, sometimes, RNAs that tell the genes when and where to turn on or off. Such *epigenetic* phenomena orchestrate the many changes through which a single fertilized egg cell turns into a complex organism. And throughout life, they enable cells to respond to environmental signals conveyed by hormones, growth factors, and other regulatory molecules without having to alter the DNA itself."[32]

The word that Dr. Pennisi uses here, *epigenetics,* is new to our lexicon. The spellchecker I am using in a 2004 version of Microsoft Word does not recognize it. The issue of the prestigious journal

Science from which her quote is taken was a special issue in 2001 devoted to the new science of epigenetics. Epigenetics, referred to by *Science* as, "the study of heritable changes in gene function that occur without a change in the DNA sequence"[33] examines the sources that *control gene expression from outside the cell.* It's a study of the signals that turn genes on and off. Some of those signals are chemical, others are electromagnetic. Some come from the environment inside the body, while others are our body's response to signals from the environment that surrounds our body.

While studying the static structure of the hard drive gives us lots of useful information, the signals that activate different sectors of the hard drive provide the source of the activation of that information. Epigenetics looks at the sources that activate gene expression or suppression, and at the energy flows that modulate the process. It traces the signals from outside the cell that tell the genes what to do and when to do it, and looks for the forces from outside the cell that orchestrate the whole. Epigenetics studies the environment, such as the signals that initiate stem cell differentiation and wound healing.

The activation of genes is intimately connected with healing and immune system function. In the studies of wound healing and marital conflict outlined above, a clear link is seen between the consciousness of the participants in the study, and the creation of the proteins (coded by gene activation) required to promote wound healing and stem cell conversion in their bodies. The healthy mental states of functional couples enabled the individuals in these relationships to emit the signals required to turn on the expression of the genes involved in immune system health and physical wound healing. Such epigenetic signals suggest a whole new avenue for catalyzing wellness in our bodies.

Magic Precedes Science

When a revolutionary new technique or therapy is described, it can take a while for science to catch up. Funding must be obtained to conduct studies. Studies must be performed, reviewed by committees of the researchers' peers, critiqued, refined, and replicated. This process takes years, and often decades. Much of the medical progress in the last fifty years has resulted from studies that build upon studies, from step-by-step incremental experimentation, with each step extending the reach of our knowledge a little bit further.

This evolutionary progress over the lifetimes of the last few generations has encouraged us to think that this is the way that science progresses. Yes, it is a way—but it is *not the only way.* There are scores of important medical procedures that were discovered years, or decades, or even centuries, before the experimental confirmation arrived to demonstrate the principles behind the treatment. Larry Dossey, in his book *Healing Beyond the Body,* urges us to "Consider many therapies that are now commonplace, such as the use of aspirin, quinine, colchicine, and penicillin. For a long time we knew that they worked before we knew how.... This should alarm no one who has even a meager understanding of how medicine has progressed through the ages."[34] "The scientist knows that in the history of ideas," observes Michael Gaugelin in *The Cosmic Clocks,* "magic always precedes science, that the intuition of phenomena anticipates their objective knowledge."[35]

The incremental approach to experimentation, with each study advancing the frontier of knowledge a little further, has served medicine well in areas such as surgery. But the incremental approach has broken down when it comes to many of the pressing afflictions rampant in our society, such as depression, Chronic Fatigue Syndrome, and autoimmune diseases. It has also made barely a dent in one of the three largest killers in Western societies: cancer. Cancer rates, when adjusted for age, have barely budged in fifty years.[36] Surgical procedures to excise cancer tumors have improved, individual drugs have been refined, and drug cocktails have been created, but these are minor variations on themes whose usefulness has been endlessly explored. Ralph Snyderman, eminent physician and researcher at Duke University, sums it up with these words: "Most of our nation's investment in health is wasted on an irrational, uncoordinated, and inefficient system that spends more than two-thirds of each dollar treating largely irreversible chronic diseases."[37]

Total health spending in the U.S. is over two trillion dollars a year; the amount spent on *all* alternative therapies is estimated at just *two tenths of one percent* of that figure.[38] For every naturopath or licensed acupuncturist in the U.S., there are seventy allopathic physicians,[39] even though such treatments can work where mainstream medicine fails,[40] are believed effective by over 74% of the population,[41] and can certainly be successful in supplementing conventional therapies.[42] It also often works better than mainstream medicine for many of the predominant disease of post-industrial cultures, such as autoimmune conditions and cancer.[43] Epigenetics gives us tools to understand why our health can be affected by so many different healing modalities.

Epigenetic Medicine

We are comfortable with incremental exploration. Yet many changes are not incremental, but very sudden. The expansion of a balloon as air is injected is smooth and incremental. A balloon popping is sudden and discontinuous. Water heated in a kettle shows little change. Then, suddenly and discontinuously, it bursts into a boil. This is the kind of breakthrough of which we find ourselves on the verge. Like the first bubbles appearing in the bottom of a pan, the possibilities of epigenetic medicine, combining integrative medicine with the breakthroughs of the new psychology, are popping through the most fundamental assumptions of our current model.

We are starting, as a society, to notice the provocative research showing the effects our thoughts and emotions have on our genes. "Science goes where you imagine it,"[44] says one researcher, and leading-edge therapies are now imagining science going in the direction of some of the powerful, safe, and effective new therapies that are emerging. Hundreds of thousands of people are dying each year, and millions more are suffering, from conditions that might be alleviated by epigenetic medicine. This book is an attempt to present this new research in a user-friendly manner that allows its power to connect with everyday experience, and to explore the potential it holds for creating massive health and social changes in our civilization in a very short time.

2

You: The Ultimate Epigenetic Engineer

We are in a school for gods, where—in slow motion—we learn the consequences of thought.

—Brugh Joy, M.D.

"Josephine Tesauro never thought she would live so long. At 92, she is straight backed, firm jawed and vibrantly healthy, living alone in an immaculate brick ranch house high on a hill near McKeesport, a Pittsburgh suburb. She works part time in a hospital gift shop and drives her 1995 white Oldsmobile Cutlass Ciera to meetings of her four bridge groups, to church and to the grocery store. She has outlived her husband, who died nine years ago, when he was 84. She has outlived her friends, and she has outlived three of her six brothers.

"Mrs. Tesauro does, however, have a living sister, an identical twin. But she and her twin are not so identical anymore. Her sister is incontinent, she has had a hip replacement, and she has a degenerative disorder that destroyed most of her vision. She also has dementia. 'She just does not comprehend,' Mrs. Tesauro says.

49

"Even researchers who study aging are fascinated by such stories. How could it be that two people with the same genes, growing up in the same family, living all their lives in the same place, could age so differently?

"The scientific view of what determines a life span or how a person ages has swung back and forth. First, a couple of decades ago, the emphasis was on environment, eating right, exercising, getting good medical care. Then the view switched to genes, the idea that you either inherit the right combination of genes that will let you eat fatty steaks and smoke cigars and live to be 100 or you do not. And the notion has stuck, so that these days, many people point to an ancestor or two who lived a long life and assume they have a genetic gift for longevity.

Josephine Tesauro and her sister

"But recent studies find that genes may not be so important in determining how long someone will live and whether a person will get some diseases—except, perhaps, in some exceptionally long-lived families. That means it is generally impossible to predict how long a person will live based on how long the person's relatives lived.

"Life spans, says James W. Vaupel, who directs the Laboratory of Survival and Longevity at the Max Planck Institute for Demographic Research in Rostock, Germany, are nothing like a trait like height, which is strongly inherited. ...'That's what the evidence shows. Even twins, identical twins, die at different times.' On average, he said, more than 10 years apart."

This report and photos, drawn from the *New York Times* in late 2006, illustrates the dramatic difference that epigenetic factors make in health and aging. Dr. Michael Rabinoff, a psychiatrist at Kaiser Permanente hospital, says that "It is known that identical twins, despite sharing the same genes, may not manifest the same psychiatric or other illness in the same way or not at all, despite the condition being thought to be highly genetic."[1] Same genes, different outcomes. Gary Marcus, Ph.D., associate professor of psychology at New York University, says it's more accurate to think of genes as "providers of opportunity" or "sources of options" than as "purveyors of commands."[2]

Think about your own life. What makes the difference between you living like Josephine Tesauro—or like her sister? Clearly, the big health differences between them can't be the result of genes, because they both started life with the same genes. It's what they did with them that counts. The epigenetic signals that make one person vibrant and the other decrepit come from outside the gene, outside the cell, and sometimes outside the body.

Cataloging the entire list of genes in the human genome is an impressive accomplishment. It's like piecing together a jigsaw puzzle of a photograph of all the members of a giant orchestra, sitting on stage, holding their instruments, ready to play. It's a static diagram of where everyone sits and what instrument they're clutching. But it tells you nothing about the choices the conductor makes for the program, about the rhythm or tone of the music, about the experience of sitting in the concert hall while a piece is being played. It tells you nothing about the swirling maelstrom of notes, what they each sound like, and how they mingle to form music. It tells you nothing about their effect on the audience. In the words of the late physicist Richard Carlson, "all the genome provides is the parts list. ...How things interact is what's more important in biology than just the things that are there. The genome tells us very little, if anything at all, about how things interact."[3] For biologists, understanding the mechanics of enormously complex self-organizing systems like the human body is a challenge of much greater magnitude than mapping the genome itself. And tracing the epigenetic influences that govern the music of the body's function is a challenge of even greater magnitude, though we see evidence of such epigenetic control every day.

To get the right answer, you have to ask the right question. Only since the concept of epigenetic control has emerged in the last decade have scientists begun to design experiments that ask these questions. As they are published, they are starting to illuminate the precise pathways by which our body takes a signal from the external environment and turns it into a set of chemical or electromagnetic instructions for our genes. One such study has gained wide attention, because it shows some of the steps required for one such interaction.

DNA is Not Destiny

One of the first animal studies that demonstrated that an epigenetic signal can affect gene expression was done with mice. While mice and humans are very different in size, they are very similar genetically, so mice are often used as subjects in laboratory experiments. In the early 1990s, researchers discovered that a gene that had long been known to affect the fur color of mice, called the Agouti gene, was related to a human gene that is expressed in cases of obesity and Type II diabetes. As well as having yellow coats, Agouti mice ate ravenously, were subject to increased incidence of cancer and diabetes,[4] and tended to die early. When they produce offspring, the baby mice are just as prone to these conditions as their progenitors.

Randy Jirtle, Ph.D., a professor of radiation oncology at Duke University, discovered, however, that he could make Agouti mice produce normal, slender, healthy young. He also discovered that he could accomplish this by changing the expression of their genes—but *without making any changes to the mouse's DNA*. This neat trick was accomplished, just before conception, by feeding Agouti mothers a diet rich in a chemical known as "methyl groups." These molecule clusters are able to inhibit the expression of genes, and sure enough, the methyl groups eventually worked their way through the mothers' metabolisms to attach to the Agouti genes of the developing embryos.

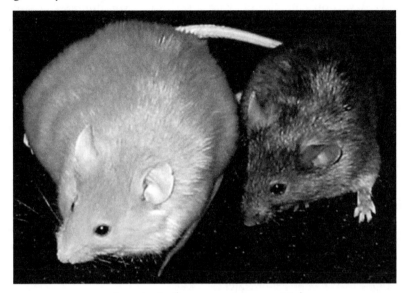

Agouti mice (gene suppression on right)

In an interview with *Discover* magazine, Jirtle said, "'It was a little eerie and a little scary to see how something as subtle as a nutritional change in the pregnant mother rat could have such a dramatic impact on the gene expression of the baby. The results showed how important epigenetic changes could be.'"[5] The article was entitled "DNA is Not Destiny: the new science of epigenetics rewrites the rules of disease, heredity, and identity." Such reports are starting to crop up in news reports with increasing frequency, as the importance of epigenetic influences becomes clearer. "The tip of the iceberg is genomics... The bottom of the iceberg is epigenetics," says Jirtle—and the larger scientific community is beginning to agree with him. In fact, in 2003, a Human Epigenome project was launched by a group of European scientists, and a U.S. project was proposed in December of 2005.[6]

Nurturing Epigenetic Change

The pathway by which epigenetic signals affect the expression of genes has many steps. Diet is the one demonstrated by the Jirtle study. A second clue comes from a series of experiments that show that *being nurtured* generates chemical changes in the brain that trigger certain genes. Dr. Moshe Szyf is a researcher at McGill University in Montreal, Canada, who studies the interactions between mother rats and their offspring. Members of his research team noticed that some rat mothers spent a lot of time licking and grooming their pups, while other mothers did not. The pups that had been groomed as infants showed marked behavioral changes as adults. They were "less fearful and better-adjusted than the offspring of neglectful mothers."[7] They then acted in similar nurturing ways toward their own offspring, producing the same epigenetic behavioral results in the next generation. This by itself is an important finding (confirmed by many other experiments) because it shows that epigenetic changes, once started in one generation, can be passed to the following generations without changes in the genes themselves.

When researchers examined the brains of these rats, they found differences, especially in a region of the brain called the hippocampus, which is involved in our response to stress. A gene that dampens our response to stress had a greater degree of expression in the well-adjusted rats.

The brains of these rats also showed higher levels of a chemical (acetyl groups) that facilitates gene expression by binding to the protein sheath around the gene, making it easier for the gene to express. Additionally, they had higher levels of an enzyme that adds acetyl groups to the protein sheath.

The anxious, fearful rats had different brain chemistry. The same gene-suppressing substance as in the Jirtle mouse study, methyl groups, was more prevalent in their hippocampi. It bonded to the DNA and inhibited the expression of the gene involved in dampening stress.

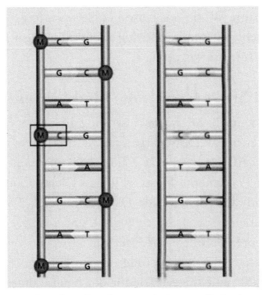

DNA strand with and without methyl group (in box) attached

To test their hypothesis that these two substances were caus-ing epigenetic behavioral changes in the rats, Dr. Szyf and his team injected the brain cavities of fearful rats with a substance that raised the number of acetyls in the hippocampus. Sure enough, the behavior of the rats changed, and they became less fearful and better adjusted. They also took the offspring of loving mothers and injected their brains with methyl groups. This produced the opposite effect; these rats became more fearful and anxious, with a heightened response to stress.

A recent article in the October/November 2006 issue of *Scientific American Mind,* notes depressed and antisocial behavior in mice, accompanied by methyl groups sticking to genes. It also extends this research to human beings; the brains of schizophrenic patients also show changes in methylation of genes, or acetylization of their protein sheaths.[8]

Mapping the protein pathways by which behaviors such as nur-turing facilitate or suppress gene expression helps us understand the implications of our behavior and beliefs, and their role in our health and longevity. The poet William Butler Yeats said, "We taste and feel and see the truth. We do not reason ourselves into it." But the huge extent to which childhood nurturing affects adult health might come as a shocking surprise to you.

Childhood Stress Results in Adult Disease

Experiments have shown a striking link between childhood stress and later disease. One large-scale, authoritative research project known as ACE, or Adverse Childhood Experiences, was done by the Kaiser Permanente Hospital in San Diego, California, in collaboration with the Centers for Disease Control. The researchers conducted detailed social, psychological, and medical examinations of 17,421 people enrolled in Kaiser's health plans over a five-year period. The study showed a strong inverse link between emotional wellbeing, health, and longevity on the one hand—and early life stress on the other. It emphasizes that there are some negative experiences that we don't just "get over," and that time does not heal.

The physicians at Kaiser scored patients on various measures of family functionality. Stressors included an alcoholic parent, divorced or separated parents, a parent who was depressed or who had a mental illness, and domestic violence. Over half the participants had experienced one or more of the defining childhood stressors, and where one stressor was present, there was an 80% chance that others were too, leading to a web of family dysfunctionality. A low score meant few stressors; a high score indicated several. The average age of study participants was fifty-seven, so in most cases it had been *fifty* years since the events occurred.

The study found that a person raised in such a family had *five times* the chance of being depressed than one raised in a functional family. Such a person was *three times* as likely to smoke. Participants who scored high on the family dysfunctionality scale were at least *thirty times more likely to attempt suicide* than those who scored low. A man with a high score was 4600% more likely to use illegal intravenous drugs. Ailments more common in those who grew up in dysfunctional families included obesity, heart disease, lung disease, diabetes, bone fractures, hypertension, and hepatitis. The genetic links between nurturing and gene expression in children is also now being traced; "one recent study suggests that children with a certain version of a gene that produces an enzyme known as MAO-A (which metabolizes neurotransmitters such as serotonin and dopamine) are significantly more likely to become violent—but only if they were mistreated as children."[9] As research proceeds, it is likely that the genetic effects of the treatment children receive will be mapped, and the epigenetic effects of parenting will be more fully understood. As a society, we will then have the understanding required to tackle social problems at their sources in childhood, rather than merely trying to ameliorate their effects played out in adulthood.

The study's authors compared our current medical practice of treating diseases to a fireman trying to get rid of billows of smoke— the most visible aspect of the problem—because of a failure to grasp that it's the underlying fire that's causing the smoke.[10] So while a study of rat pups might seem like an Ivory Tower exercise in epigenetics, the reality of nurturing in the real world makes a difference in the health and wellbeing of millions of people.

It's the Gene Show, and You're the Director

There are certainly lifestyle factors that make a big difference in our health and longevity. Having a Body Mass Index of twenty-five or less, eating a diet rich in fruits and vegetables, daily aerobic exercise, avoiding smoking and excess alcohol—all these contribute to living to a ripe old age. There may be an epigenetic component to each of them too. Yet there is mounting evidence that invisible factors of consciousness and intention—such as our beliefs, feelings, prayers, and attitudes—play an important role in the epigenetic control of genes.

The old view that our genes contain indelible instructions governing the functioning of our bodies is, in the scornful words of my offspring, "*So* twentieth century." We now understand that a host of other factors determine which genes are expressed. Some of these are physical, like diet, exercise, and lifestyle. Others are metaphysical, like beliefs, attitudes, spirituality, and thoughts. It's taken science a long time to figure out that something as seemingly immaterial as a belief can take on a physical existence as positive or negative changes in our cells. But it turns out that these factors can affect health and longevity dramatically. Josephine Tesauro and her sister were born with an identical collection of instruments in their genes. The music they played in their first years may have been indistinguishable. But the finale of each of their life concerts is quite different.

As we hold the scale of health in our hands, with good health on one side and decrepitude on the other, we can tilt the outcome. If we can add a brick to the side of good health, we can tilt it in our favor. Let's take a look at some of the bricks we can drop on our scale. Each of these is based on sound scientific research and holds lessons we can apply from this day forward.

Beliefs and Biochemistry

A landmark study linking belief to health was reported recently by Gail Ironson, M.D., Ph.D., a leading mind-body medicine researcher, and Professor of Psychology and Psychiatry at the University of Miami. Dr. Ironson runs the Positive Survivors Research Center at the university, and has been awarded several grants from the National Institutes of Health. It is one of the first studies to link particular beliefs with particular changes in the immune system.[11]

Dr. Ironson measured several indicators of health in HIV patients over the course of a four-year period. One measure was their viral load—the quantity of the AIDS virus in a sample of blood. She also counted the concentration of a type of white blood cell responsible for killing invading organisms. The concentration of these "helper T-cells" (also known as CD4 cells) in the blood is one measure of the progression of AIDS. If the concentration of helper T-cells drops, our bodies are less able to fend off other diseases like pneumonia. That's why the "I" and "D" in AIDS stand for Immune Deficiency; as AIDS patients lose their T-cells and their immunity to disease drops, they are more susceptible to the kinds of invading organisms—opportunistic infections—that healthy immune systems easily fend off.

Studies like those conducted by Dr. Ironson are especially meaningful to physicians and biologists because they identify key *biological* markers of illness, as opposed to *subjective* measures such as the patient's level of depression, the number of doctor visits, and the dosage of medication required.

In her studies, Dr. Ironson found that there were two particularly interesting predictors of how fast HIV progressed in the bodies of her research participants. The first was their view of the nature of God. Some believed in a punishing God, while others believed in a benevolent God. She observes that, "People who view God as judgmental God have a CD4 (T-helper) cell decline more than twice the rate of those who don't see God as judgmental, and their viral load increases more than three times faster. For example, a precise statement affirmed by these patients is 'God will judge me harshly one day.' This one item is related to an increased likelihood that the patient will develop an opportunistic infection or die. These beliefs predict disease progression even more strongly than depression."

Dr. Ironson was surprised to find that many people reported a spiritual transformation subsequent to their diagnosis. This transformation was characterized by a sense of self that was profoundly changed, and resulted in different behaviors. Many kicked their habits of street drugs like cocaine and heroin, or legal ones like alcohol. Some went through such a transformation only after hitting rock bottom. Carlos, one of them, describes his experience of getting to the end of his rope:

> I was planning to finish my BA, moving to New York. I found out that my ex partner had been doing drugs and cheating with other relationships. I was very scared, and I didn't deal with it. For six months I didn't get tested. When I did find out, I had no friends in New York so I had to deal with it on my own. I turned to cocaine, my life changed dramatically; I was sort of spiraling down hill, near the lowest point in my life. It changed

everything, it changed my behavior, it changed my ambition, I didn't have the same drive that I had going in after school to pursue my career. Things were so bad that any belief that I had in a higher being or in a spiritual presence was completely extinguished. I was on a course down hill. I just didn't care.[12]

After being diagnosed as HIV positive, Carlos's infection progressed rapidly into full-blown AIDS. He suffered from serious opportunistic infections, and had very low levels of T-cells, and high levels of viral load, despite taking HIV medication.

A common gateway to spiritual transformation was having a spiritual experience. After helping a drunk white man in distress, John, a gay African-American man with a college education, described the following experience:

> I felt like I was floating over my body, and I'll never forget this, as I was floating over my body, I looked down, it was like this shriveled up prune, nothing but a prune, like an old dried skin. And my soul, my spirit was over my body. Everything was so separated. I was just feeling like I was in different dimensions, I felt it in my body like a gush of wind blows. I remember saying to god, "God! I can't die now, because I haven't fulfilled my purpose," and, just as I said that, the spirit and the body, became one, it all collided, and I could feel this gush of wind and I was a whole person again.
>
> That was really a groundbreaking experience. Before becoming HIV-positive my faith was so fear based. I always wanted to feel I belonged somewhere, that I fit in, or that I was loved. What helped me to overcome the fear of God and the fear of change was that I realized that no one had a monopoly on God. I was able to begin to replace a lot of destructive behavior with a sort of spiritual desire. I think also what changed, my desire to get close to God, to love myself, and to really embrace unconditional love.[13]

John's story points to the second major factor Dr. Ironson noted: A participant's personal relationship with God. Her study found that patients who did not believe that God loved them lost helper T-cells "three times faster than those who believed God did love them."[14] Another correlation she found was that those who felt a sense of peace also had lower levels of the body-damaging stress hormone cortisol.[15]

Dr. Ironson, in her recent article published in the *Journal of General Internal Medicine,* showed a fairly high number of people increase their spirituality in the year after they are first diagnosed with HIV/AIDS. 45% showed an increase in spirituality, 42% stayed the same, and 13% had a decrease in spirituality. The study showed an enormously strong association between spirituality and the progression of HIV.[16]

"I was surprised that so many people had an increase in spirituality, because being diagnosed with HIV/AIDS can be a devastating event. I could hardly believe the figures, until I saw that another article in the same issue of the journal found an increase in spirituality of 41% of newly diagnosed patients. Perhaps a life-threatening illness, not just HIV, but cancer or a heart attack, can stimulate a person to reexamine their connection to the sacred."

Dr Ironson summarizes by saying that, "If you believe God loves you, it's an enormously protective factor, even more protective than scoring low for depression, or high for optimism. A view of a benevolent God is protective, but scoring high on the *personalized* statement 'God loves me' is even stronger."[17]

This echoes another study that found that, "Patients who believed that God was punishing them, didn't love them, didn't have the power to help, or felt their church had deserted them, experience 19% to 28% greater mortality during the 2-year period following hospital discharge."[18]

Unfortunately, many more Americans believe in the God of thunderbolts and retribution than believe in a benevolent God. In a study done by Baylor University's Institute for Studies in Religion, researchers found that 31% of Americans see God that way. The number of people believing in Authoritarian God goes high as 44% of the population in the country's Southern states.

Just 23% of the population believes in Benevolent God, according to the study, while the rest fall in the middle. They believe in a Critical God (16%), Distant God (24%), or are atheists (5%).[19] Since our view of God can have such huge effects on our health, it's worth examining our beliefs, and if our religion or spiritual orientation permit such recalibration, adjusting them to fit the most loving vision of God of which we are capable. Carlos, the young man who hit bottom in Ironson's HIV/Spirituality study,[20] says,

> You don't have to believe in any God that doesn't love you or any God that isn't here to help you. Because I had a Catholic background, during my addiction I felt like I was being judged, that I was being punished. I thought I was going to die for my sins. So when I went to this service and I heard [the minister talk about choosing a loving God, it] changed my God to one that was loving and helpful. It was revolutionary.

Shortly thereafter, Carlos went to Alcoholics Anonymous and became sober. And while you and I are probably not in the same dire straits as he was, our bodies will be deeply grateful for us having enough faith in them to adjust our religious faith in the direction of a loving God.

Psychology Becomes Physiology

What we believe about what is happening to us enhances the facts. A 2007 Harvard study examined the difference between physical exertion, and physical exertion plus belief. The researchers recruited eighty-four maids who cleaned rooms in hotels. The sample was divided into two groups. One group heard a brief presentation explaining that their work qualifies as good exercise. The other group did not.

Over the next thirty days, the changes in the bodies of the women who had heard the presentation were significant: "The exercise-informed women perceived themselves to be getting markedly more exercise than they had indicated before the presentation. Members of that group lost an average of 2 pounds, lowered their blood pressure by almost 10 percent, and displayed drops in body-fat percentage, body mass index, and waist-to-hip ratio."[21]

This marked physiological change occurred in just thirty days, and followed one brief session in which the researchers exposed the women to new beliefs about their level of physical activity. Imagine the effect of the background music of our own self-talk, running in a continuous loop in our heads for many hours a day, as we perform our daily routines. Making even small changes in the program can lead to significant changes in our health.

Prayer

Prayer is one of the most powerful forms in which intention may be packaged. Prayer has been the subject of hundreds of studies, most of which have demonstrated that patients who are prayed for get better faster.

One such study was done by Thomas Oxman and his colleagues at the University of Texas Medical School. It examined the effects of social support and spiritual practice on patients undergoing heart surgery. It found that those with large amounts of both factors exhibited a mortality rate *just one-seventh* of those who did not.[22] Another was done at St. Luke's Medical Center in Chicago. It examined links between church attendance and physical health. The researchers found that patients who attended church regularly, and had a strong faith practice, were less likely to die and had stronger overall health.[23]

These are not isolated examples. Larry Dossey, in *Prayer is Good Medicine,* says that there are over 1,200 scientific studies demonstrating the link between prayer and intention, and health and longevity. Meta-analyses in the *Annals of Internal Medicine*[24] and *The Journal of*

Alternative and Complementary Medicine[25] have compiled the results of many studies and found that prayer, distant healing, and intentionality have significant effects on healing.

Even a recent confounding study published in the *American Heart Journal* tells us more about the limits of scientists' understanding of prayer than it tells us about prayer itself. Under headlines like the one in the *Washington Post* on March 31, 2006: "Strangers' prayers didn't help heart patients heal," stories about this large-scale study of 1,800 patients undergoing heart bypass surgery reported that those who were prayed for had as many complications as those who did not.

A variety of explanations were advanced for why prayer had appeared to fail in this study, and some scientists opposed to prayer studies argued that it was so conclusive that further money should not be put into a consciousness-based intervention that had been so thoroughly debunked.

I was most surprised at the study's results, until I read the fine print. It turned out that, in order to "standardize" what was meant by prayer, the researchers had designed the study so that patients were prayed for only starting the day of surgery (or the evening before), and continuing for fourteen days afterward. In addition, a standard eleven-word prayer was used for every patient: "For successful surgery with a quick healthy recovery and no complications."[26]

Such sterilized prayer cannot be as successful as heartfelt, spontaneous prayer. Other studies have shown that the skill and fervor of the person praying has a marked effect on the subject of prayer. One controlled, randomized, double-blind study reported in Dossey's *Prayer is Good Medicine* measured the ability of people to increase the growth of yeast in test tubes. Three of the people were healers (one an M.D. who practiced spiritual healing) and the other four were student assistants. The results showed that mental concentration and intention definitely affected the growth of the yeast. "Analysis revealed that there were fewer than two chances in a hundred that the positive results could be obtained by chance. The bulk of the positive scores was credited to the three healers. When their scores were analyzed separately, there were fewer than four chances in ten thousand that the results could be due to chance..."

In a careful study of distant healing prayer, the healers used their own unique methods, which ranged from putting photos of the patient on an altar with a statue of the Virgin Mary, to Sioux peace pipe ceremonies, to the "projection of qi."[27] Also, the healers repeated their intentions daily for ten weeks. In other tests in which prayers sought to influence the germination rate of seedlings, "the more experienced practitioners produced the more powerful outcomes. These studies indicate that practice, interest, and experience make a difference in spiritual healing, which for most healers is based in prayer."[28]

The failure of the cardiac prayer study to show an improvement was due, I believe, to the scripted and structured nature of the "prayer" designed so carefully by the researchers, but which squeezed out any fervor, passionate intent, or personalization by the person doing the praying. To be powerful, intent must be deeply, personally, and sincerely engaged. The researchers in the cardiac study were not studying the effects of prayer: they were studying the effects of their own belief of what prayer comprises.

Doing Good Does You Good

Besides helping the person prayed for, it is likely that prayer benefits the person doing the praying. Studies show that regular acts of altruism prolong our lives and improve our own happiness.[29] Prayer is good medicine for the person doing the praying as well as the receiver.

In her book *The Energy Prescription*, pharmacist Constance Grauds, R.Ph., describes one such study done in Michigan. It included a large sample, 2,700 men, and it studied them over a long period—ten years. It found that the men who engaged in regular volunteer activities had death rates half of those who did not. She says that, "altruistic side effects include reduced stress; improved immune system functioning; a sense of joy, peace, and wellbeing; and even relief from physical and emotional pain. These effects tend to last long after the helping encounter, and...increase with the frequency of altruistic behavior."[30]

Seven Minutes of Spirituality

A study that demonstrates the effect of spiritual nurturing was performed by Jean Kristeller, Ph.D., a psychologist at Indiana State University. She reported that when doctors spent time talking with critically ill cancer patients about their *spiritual* concerns, follow up revealed that after three weeks, the patients reported a better quality of life and less incidence of depression. Patients who had been talked to also felt that "their physicians cared more about their health, which was in contrast to those patients in the study whose physicians did not discuss spiritual matters with them."[31]

And the length of time of the discussion that so affected patients' lives for weeks afterwards? A mere *five to seven minutes!*

Meditation

The benefits of meditation are so numerous, and the subject of so many studies,[32] that it's hard to know where to start. Dr. Robert Dozor, co-founder of the Integrative Health Clinic of Santa Rosa, California, says, "Meditation—all by itself—may offer more to the health of a modern American than all the pharmaceutical remedies put together."[33] Recently, neuroscientist Richard Davidson, Ph.D., of the University of Madison at Wisconsin, has published a series of experiments using PET scans and EEG recordings to study the areas of the brain that are active during meditation.

When comparing the results obtained by novice meditators against those of experienced meditators such as Tibetan Buddhist monks, it was found that the monks, "showed greater increases in gamma waves, the type involved in attention, memory, and learning, and they had more brain activity in areas linked to positive emotions like happiness. Monks who had spent the most years meditating had the greatest brain changes."[34] This means that we are bulking up the portions of our brains that produce happiness when we meditate. Another report noted that, "In a pilot study at the University of California at San Francisco, researchers found that schoolteachers briefly trained in Buddhist techniques and who meditated less than 30 minutes a day improved their moods as much as if they had taken antidepressants."[35] Love and compassion are health-skills in which we can train ourselves.

Epigenetic Visualizations

The use of visualizations to help patients cope with cancer was pioneered by Carl Simonton and others in the 1970s. I vividly remember an interview I did with a woman in 1989. She impressed me as someone with great strength of will and courage.

Nancy had been diagnosed with metastacized Stage IV uterine cancer in 1972. Though her condition was terminal, she had rejected conventional medical therapy entirely, reasoning, "My body created this condition, so has the power to uncreate it too!" She quit work, exercised as much as her physical energy allowed, and spent hours lying in the bath. She came up with a visualization that tiny stars were coursing through her body. Whenever the sharp edge of a star touched a cancer cell, she imagined it puncturing the cancer cell, and the cancer cell deflating like a balloon. She imagined the water washing away the remains of the dying cancer cells. She focused on what she ate, how far she could walk, her baths, and the stars, and little else.

Nancy began to feel stronger, and her walks became longer. She began to visualize what her future might look like many years from that time. She went back to see her doctor three months after the diagnosis. She did not make the appointment until she had a firm inner conviction that the cancer was completely gone. To the astonishment of her physicians, tests revealed her to be cancer-free. Curiously, many patients who use similar techniques report an inner knowing that the disease is gone, long before it is confirmed by medical tests.[36] They also use highly individualized images that work for their particular psyche.

Many years later, Nancy was still in excellent health, and she would occasionally still visualize the stars rushing through her body, carrying away whatever traces of cancer might still remain.

It's that last detail that points to the preventive possibilities in epigenetic medicine. Meta-analysis of large bodies of research indicates that many genes express differently in cancer patients than they do in people without cancer.[37]

It's possible that Nancy's ongoing "star-cleaning" visualizations, long after she was diagnosed as cancer-free, helped keep her genetic profile favorable to cancer remission. Such visualizations are also free, safe, and non-invasive. Their ongoing effectiveness could be verified with DNA screening, biomarkers, and other non-intrusive tests.

The possibilities of visualization for epigenetic healing are indicated by a recent study that examined how the expectations of seventh grade students affected their math scores. Stanford University research psychologist Carol Dweck, Ph.D. noticed that students had beliefs about the nature of intelligence, and it had an effect on their performance. Some students believed that intelligence is a fixed quantum, like the number of inches in your height, or the number of teeth in your mouth. Others believed that intelligence can grow and develop, like a plant. She then compared the math scores of the two groups over the course of the following two years.

She found that students who believed that intelligence can grow had increasing math scores. The math scores of those who believed that intelligence is fixed decreased.

Dweck then wondered, "If we gave students a growth mindset, if we taught them how to think about their intelligence, would that benefit their grades?" She took a group of one hundred seventh graders who were all performing badly in math and divided them, at random, into two groups. The first group received instruction in good study skills. The second group received information about the ways our brains grow and form new neural connections when confronted with novelty and challenge. They, "learned that the brain actually forms new connections every time you learn something new, and that, over time, this makes you smarter." At the end of the semester, those students who had received the mini-course in neuroscience had significantly better math grades than the other group. Dweck says, "When they worked hard in school, they actually visualized how their brain was growing."[38] This visualization had concrete effects on their academic performance. It's not unreasonable to assume that visualization can have the effect of increasing health. What we imagine, we can create. Filling our minds with positive images of wellbeing can produce an epigenetic environment that reinforces the healing process.

Attitude Is Everything

"Attitude is everything with aging," says Dr. Andrew Weil, author of *Spontaneous Remission* and several other books. He cites studies that show that negative perceptions about aging can shorten our lives, while positive beliefs prolongs them: older people with positive attitudes about aging were found to live 7½ years longer than those with negative attitudes. He also reminds us that optimism heals: "A study of nearly 1,000 older adults followed for nine years concluded that people with high levels of optimism had a 23% lower risk of death from cardiovascular disease and a 55% lower risk of death from all causes compared to their more pessimistic peers." Positive older people also have better memories and stay healthier. Overall physical fitness is reflected in walking speed; positive elders were found to walk 9% faster than negative ones.[39]

Neurosurgeon Norman Shealy, M.D., Ph.D., in his book *Life Beyond 100,* summarizes four personality types and—based on many studies—links them to longevity. The first type has a lifelong pattern of hopelessness. The second group has a lifelong pattern of blame or anger. The third group bounces between hopelessness and anger. And the fourth group is self-actualized. They believe that "happiness is an inside job."

He reports that people in the fourth category tend to die of old age, and that less than 1% of people in this category die of cancer or heart disease. About 9% of people in the third group die of one of those two conditions.

By way of contrast, he finds that 75% of people who die of heart disease, and 15% of those who die of cancer are members of the Lifelong Anger Club, group two. And group one, those with lifelong patterns of hopelessness, tend to die thirty-five years younger than those in group four. Seventy-five percent of them die of cancer, and 15% of heart disease.[40]

While attitudes such as optimism and positivity were once regarded as accidents, research like that of Richard Davidson is demonstrating that they are also learned skills. They can be cultivated. Knowing that we are having powerful genetic effects on the production of healing proteins in our bodies provides a strong incentive to learn techniques for improving our attitudes, a therapeutic tool that can exceed the promise of most conventional therapies. As you contemplate the fork in the road between positive and negative attitudes, imagine yourself splitting into two genetically identical individuals. Both are you at the present moment. Then fast forward twenty years. Imagine that one of the twins has taken conscious control of attitude, and the other has not. Which one would you rather be?

Why Stress Hurts

What you are thinking, feeling, and believing is changing the genetic expression and chemical composition of your body on a moment-by-moment basis. The stress hormone cortisol has the same chemical precursor as DHEA, which is associated with many protective and health-promoting functions, and contributes to longevity. When that precursor is being used to make cortisol, production shifts away from making DHEA. When our cortisol levels are low, the raw materials from which our bodies manufacture life-giving DHEA are freed up, and production of DHEA increases. As one researcher puts it: "When our energy reserves are continually channeled into the stress pathway, there isn't enough energy left to support regenerative processes that replenish the resources we've lost, repair damage to our bodies, or protect us against disease. ...The repair and replacement of most kinds of cells is diminished; bone repair and wound healing is slowed, and levels of circulating immune cells and antibodies fall. ...In high levels the stress hormone cortisol kills our brain cells."[41] Cortisol has been shown to reduce muscle mass, increase bone loss and osteoporosis, interfere with the generation of new skin cells, increase fat accumulation around the waist and hips, and reduce memory and learning abilities.[42] Low levels of DHEA have been linked to a multitude of diseases.

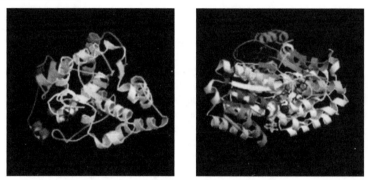

DHEA (left) and cortisol (right) are both manufactured by the adrenal gland using the same precursors

Engineering Your Cells Consciously

The body's stress response encompasses far more than shunting production away from DHEA to produce cortisol. Over 1,400 chemical reactions and over thirty hormones and neurotransmitters shift in response to stressful stimuli. So by de-stressing ourselves using attitude, belief, nurturing, self-talk, and spirituality, we are taking a role in determining which instruments in our genetic symphony predominate. This knowledge opens up a panorama for self-healing as vast as the number of moments left in your life. When you understand that *with every feeling and thought, in every instant, you are performing epigenetic engineering on your own cells,* you suddenly have a degree of leverage over your health and happiness that makes all the difference. How you use that knowledge can determine whether your unique symphony comes to an early and discordant end, or whether you play beautiful music to a long finale.

When you choose beliefs, feelings and other epigenetic influences that benefit your health, you can create a virtuous cycle of epigenetic health. In an epigenetic health cycle, you intervene consciously with positive emotions, thoughts and prayers. Besides making you feel good psychologically, these benefit your body, modulating your gene expression in the direction of the highest peak of health available to you.

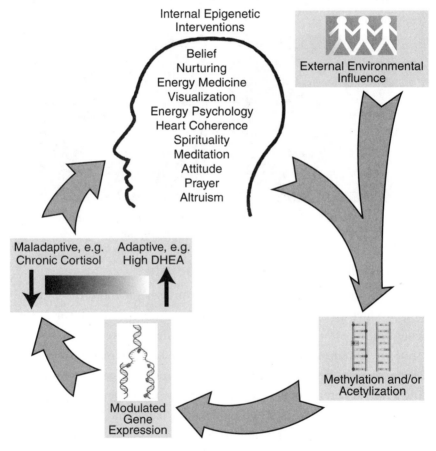

Epigenetic health cycle

This peak of health differs from person to person, and there is little value in looking at your personal peak and comparing it to that of someone else, even an identical twin. For there are external influences that are beyond our control, and they can have profound epigenetic impact: Consider, for instance, two identical twins, one of which receives more nurturing than the other. Or think about twins living in different towns, one of which has severe environmental pollution while the other does not. Like the car that rams into you from behind on the freeway, there are random life situations that are beyond our control.

Yet no matter how well or sick we may currently be, we still have the ability to choose our thoughts and feelings, and select those that support peak vitality. I call this the *epigenetic health cycle.*

In an epigenetic health cycle, we select positive beliefs, prayers and visualizations that support peak health. We avoid those that do not. In this way, we consciously intervene to send epigenetic signals to our cells. These signals can reduce stress, and promote the synthesis of life-enhancing hormones like DHEA, as well as thousands of other beneficial substances.

We've seen how powerful each of these little bricks can be in tipping the scale of our health. Positive self-talk, nurturing beliefs, altruism, attitude, meditation, and prayer can add brick after brick to the scale. But what if we had at our disposal truckload of bricks to dump on the side of good health? Some of the emerging new therapies promise just this kind of decisive intervention, as we will see in the coming pages.

3

The Malleable Genome

The longest journey is the journey inward, for he who has chosen his destiny has started upon his quest for the source of his being.
—Dag Hammarsjkold, former United Nations Secretary-General

"A therapist[1] who was having her home remodeled noticed that one of the construction workers had a strange skin condition on his arms. She asked him about it, and he said, 'Oh, that's my psoriasis. Had it for years.' He turned his arms around to show her that from his wrists to his shoulders, his skin was a bubbling ocean of peeling skin with sore red tissue and fluid beneath it. She replied, 'Ooh, that must be painful. How did you get it?' 'Well I don't know,' he replied, 'I guess it started about three years ago, when my girlfriend told me she was pregnant.' The therapist asked if he might like to try a new treatment she knew. He was dubious. His doctor had told him that there was nothing to be done. The therapist replied, 'Ah, yes, what he meant was, there's nothing to be done with pills, ointments, and injections. Your mind created this, and so only your mind can take it away again.' He nodded and she showed him just a bare-bones procedure, suggesting that three times each day he do three rounds of tapping to the statement 'I want to get over my psoriasis.' Within two weeks, his skin had healed on both arms, down to a small patch the size of a coin on his elbows. In addition, he has since used the approach to overcome lower-back pain that had troubled him for years."[2]

That these kinds of shifts are possible with energy medicine and Energy Psychology is being established by a rapidly growing database of research studies and clinical case histories. What have not been at all clear up to this point are the mechanisms behind such shifts. Science is never content with knowing *what;* it rightly insists on knowing *why.* While the "what" has been accumulating in large stacks, precise experimental descriptions of the "why" have lagged behind until the right questions about epigenetic control were asked and answered.

This book could not have been written ten years ago, because there were not yet enough credible and well-designed scientific studies to support its hypotheses. Today, there are, and the number of published experiments will grow exponentially in the coming decade. With a catalog of genes in place, researchers are now focusing on how these genes work. In some cases they work singly, but most often they work in concert, both with each other, and with signals from the inner and outer environments.

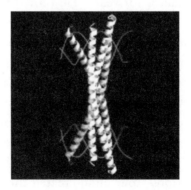

X-ray diffraction image of human DNA bound in protein

Very few human processes are turned on or off by a single gene. Most processes require many genes, acting together to produce a common result. The idea, fostered by the mass media, that there is a gene for this or a gene for that, is incorrect. Genes are implicated in conditions in a variety of ways. Headlines like the October 29, 2005 proclamation in the *New York Times* stating, "Two More Genes Linked to Dyslexia" (in addition to a third gene announced a year earlier) over-simplify the cascade of genetic factors involved in conditions. After a subtitle that says, "Findings Support that Disorder is Genetic," the story goes on to tell us that, "...people deemed simply lazy or stupid because of their severe reading problems may instead have a genetic disorder that interfered with the wiring of their brains before birth. 'I am ecstatic about this research,' said Dr. Albert Galaburda of Harvard Medical School, a leading authority on developmental disorders..."

Such "breakthroughs" create elation that is often followed by disappointment, as the complexity of the genetic interactive systems later becomes clear. For example, more sober reports have implicated some 600 genes that express differently in patients with heart disease. Researchers have shown that hundreds of genes are implicated in certain other diseases for which the genetic profile has been mapped.

As well as *many* genes being involved in most changes of state, different genes are often involved at *different time periods* of that change of state. Not just from day to day, but from second to second, genetic cascades are turned on or off by our experience. Some genes may engage early, others may express afterwards, while yet others may reach peak expression many hours later, in a complex and coordinated dance. There are several ways of profiling genes, and one way they may be cataloged is to look at the *speed at which they reach peak expression* when stimulated by an environmental influence.

Some genes are activated quickly; others more slowly. Genes that activate very quickly (sometimes *within one or two seconds* of a stimulus) are called *immediate early genes*. Their function is often to trigger the activation of other genes. In general, *early activated genes* reach their peak of expression in roughly an hour. A second class of genes, the *intermediate genes,* reach their peak of expression in about two hours. *Late genes* take longer periods to reach their peaks, up to eight hours, and their effects may last for a period of a few hours, or for a period of years. Certain classes of late activated genes, once expressed, may remain "on" for your entire lifetime. An early firing gene may have, as a primary purpose, the activation of several intermediate and late-firing genes.

Besides being identified with the speed at which they express, genes may be classified by the kinds of triggers that turn them on and off. One such class of genes is called *experience-dependent genes* or activity-dependent genes. These genes are involved in activities or experiences such as growth, healing, and learning. Another class of genes we will examine in more detail is *behavioral state-dependent genes.* These genes come into expression during periods of stress or emotional arousal, or in different states of awareness such as dreaming and dreamless sleep.

Gene Chips

How do we know which DNA molecules are expressed in, for instance, a blood sample from a cancer patient? One of the newest tools that has enabled researchers to conduct experiments that show particular genes being triggered is the DNA microarray. Such gene chips assemble thousands of different strands of DNA onto a single wafer. When exposed to a sample, they can then demonstrate which of the strands have been affected by the sample.

The characteristic of DNA that allows identification of expressed genes is simple in principle. The double helix shape of a DNA molecule looks like a spiraling zipper. During replication, the molecule becomes partially "unzipped." The two halves of the zipper separate, and each seeks another half-molecule to bond with. While it is unzipped and separated, bonds on the unzipped portion seek other interlocking half-zippers to attach to. Gene chips contain hundreds

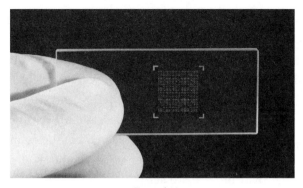

Gene chip

or even thousands of tiny wells, like pixels on an LCD screen. To the bottom of each of these wells, a particular unzipped strand is attached during the process of manufacturing the chip. These "sticky" half-molecules seek their counterparts, and by examination of which ones are able to bond, researchers can identify exactly which DNA strands in a sample are active.

A gene chip yields results that are also somewhat like an LCD screen with thousands of pixels. Each pixel is a different color. Researchers can note which pixels change color when exposed to a sample, and are thus able to "read" the gene chip to find out which molecules have been able to bond with a counterpart in the sample. This sophisticated technology allows researchers to identify DNA states, and changes in DNA, in a variety of conditions.

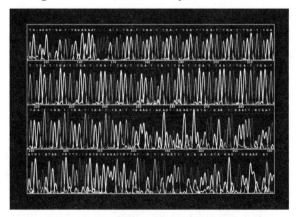

Portion of a visual map derived from a gene chip,
showing the expression of various genes[4]

Gene chips will be put to many more experimental uses in the years to come, giving researchers much clearer ideas about what happens to genes under a variety of conditions. They have already been used to study the effects of acupuncture.[5] It will be especially interesting to find out what gene changes occur before and after meditation, prayer, therapeutic touch, energy therapies, and other consciousness-based treatments. Links to relevant research can be found at www.EpigeneticMedicine.org, and on other research and medical sites. At Soul Medicine Institute, we are actively encouraging research on the protein and genetic components of changes in consciousness such as intention, belief, altruism, and prayer (see Appendix E for details of these experiments).

Immediate Early Genes

This class of genes responds within minutes to events that happen in our lives, and to cues from our environments. They mediate between the environment and the body's neurochemical processes. They activate other genes, which in turn code for the proteins that govern our cells' ability to adapt to health and illness. Early activated genes reach their peak expression in around one hour, while the peak is about two hours for intermediate activated genes. But many immediate early genes express in much shorter time periods—between *two seconds* and *two minutes.*

A class of immediate early genes regulates our body's wakefulness and sleep. They are the chronobiological or "clock genes." Researcher Marina Bentivoglio, of the Department of Neuroscience at Stockholm's prestigious Karolinska Institute, says that "The study of immediate early genes indicates that sleep and wake, as well as synchronized and desynchronized sleep, are characterized by different genomic expressions, the level of IEGs being high during wake and low during sleep. Such fluctuation of gene expression is not ubiquitous but occurs in certain cell populations of the brain."[6] Just as certain areas of the brain show increased activity of certain waveforms during certain states, they also show increased gene expression. Immediate early genes can be activated both by cognitive changes in the individual, and also by cues such as sexual stimulation from outside.[7]

An example of the pathway followed in such transformations is demonstrated by an immediate early gene called *C-fos*. It is part of a class of immediate early genes that modulates our body's response to stress; this class is activated by stressful situations—whether they're interoffice rivalries, marital disagreements, or attacks by wild animals. C-fos activates brain neurons to produce a protein called *fos*. Fos then binds to the DNA molecule where it triggers the transcription of other genes. The stresses that trigger the activation of C-fos can be physical traumas. They can also be stressful social or psychological situations. In this way, *this family of immediate early genes sets up the response of the rest of the body's mechanism for dealing with stress.*

Gary Marcus, Ph.D, author of *The Birth of the Mind: How a Tiny Number of Genes Creates the Complexity of Human Thought,* says that, "A single regulatory gene at the top of a complex network can indirectly launch a cascade of hundreds or thousands of other genes," and that by, "by compounding and coordinating their effects, genes can exert enormous influence on biological structure." He gives examples of experiments in which "a simple regulatory gene leads directly and indirectly to the expression of approximately 2,500 other genes."[8]

Positive stress can optimize the functioning of our immune systems. Hans Selye, the originator of the concept of stress, originally broke it down into two poles: *distress,* or negative stress, and *eustress,* or positive stress—the kind of stress that causes an athlete to excel or a painter to reach new heights of inspired creative expression. Unfortunately, in keeping with human experiences, the word *distress* has endured, and the word *stress* has become associated exclusively with distress, while the word *eustress* never gained wide acceptance.

Immediate early genes are also critical to the functioning of our immune systems. Distress, whether sourced from within or without the individual, can depress the expression of genes that enhance the functioning of our immune system. With medical students in the midst of their final exams as his subjects, researcher Ronald Glaser studied the effect of stress on one of the immune system's messenger molecules. The molecule, interleukin-2, instructs helper T-cells— white blood cells that devour diseased cells and intruders—to attack. He found that during this stressful period, the students showed a significant drop in interleukin-2 production,[9] implying a corresponding drop in the transcription of the gene that regulates interleukin-2 production. In follow-up studies, Glaser also found that the stress pre-cancerous subjects were experiencing led to reduced expression of two immediate early genes associated with immune function: c-myc and c-myb.[10]

Conversely, positive influences—eustress—can bolster the genetic component of our immune systems. Immunologist M. Castes, Ph.D., showed that emotionally supportive experiences of children in therapeutic support groups improved aspects of their immune system function that depend on genetic activation. The group in his experiment went through a six-month program of self-esteem workshops, guided imagery, and relaxation. When compared with a control group that had not had the same environmental stimuli, the children in the experimental group had both fewer episodes of asthma and fewer incidences of the use of anti-asthma medication. Immune factors in the blood of the experimental group increased, as did gene expression of the factors governing interleukin-2 production.[11]

Immediate early genes can also affect the developing fetus. Some of the genes activated by immediate early genes shape the form and functioning of the body. When stress proteins are present during fetal development, they can shape the anatomy or rate of growth[12] of the child; "The stress may be momentary, and the arousal of the early genes brief. Yet the effects triggered by the activation of the genes they act upon may produce long-term changes."[13]

Behavioral State-Related Genes

During various states of awareness, like sleeping and waking, strong emotional arousal, distress and eustress, different patterns of genes express. These genes are related to our behavioral state, and

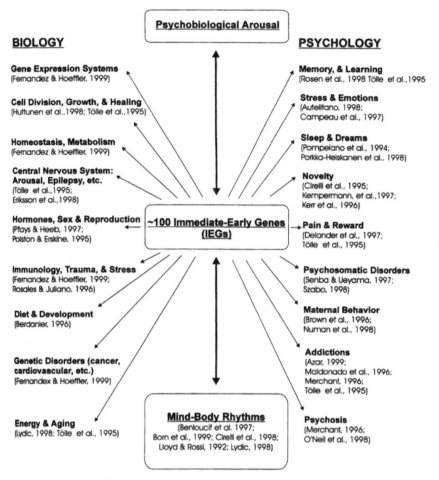

BIOLOGY

Gene Expression Systems
(Fernandez & Hoeffler, 1999)

Cell Division, Growth, & Healing
(Huttunen et al.,1998; Tölle et al.,1995)

Homeostasis, Metabolism
(Fernandez & Hoeffler, 1999)

Central Nervous System: Arousal, Epilepsy, etc.
(Tölle et al.,1995;
Eriksson et al.,1998)

Hormones, Sex & Reproduction
(Pfays & Heeb, 1997;
Polston & Erskine, 1995)

Immunology, Trauma, & Stress
(Fernandez & Hoeffler, 1999;
Rosales & Juliano, 1996)

Diet & Development
(Berdanier, 1996)

Genetic Disorders (cancer, cardiovascular, etc.)
(Fernandex & Hoeffler, 1999)

Energy & Aging
(Lydic, 1998; Tölle et al., 1995)

Psychobiological Arousal

~100 Immediate-Early Genes (IEGs)

Mind-Body Rhythms
(Benloucif et al. 1997;
Born et al., 1999; Cirelli et al., 1998;
Lloyd & Rossi, 1992; Lydic, 1998)

PSYCHOLOGY

Memory, & Learning
(Rosen et al. 1998 Tölle et al.,1995

Stress & Emotions
(Autelitano, 1998;
Campeau et al., 1997)

Sleep & Dreams
(Pompeiano et al., 1994;
Porkka-Heiskanen et al., 1998)

Novelty
(Cirelli et al., 1995;
Kempermann, et al.,1997;
Kerr et al., 1996)

Pain & Reward
(Delander et al., 1997;
Tölle et al., 1995)

Psychosomatic Disorders
(Senba & Ueyama, 1997;
Szabo, 1998)

Maternal Behavior
(Brown et al., 1996;
Numan et al., 1998)

Addictions
(Azar, 1999;
Maldonado et al., 1996;
Merchant, 1996;
Tölle et al., 1995)

Psychosis
(Merchant, 1996;
O'Neil et al., 1998)

*Immediate early genes play a vital role in regulating a great
many psychological and physiological functions*[14]

are therefore known as behavioral state-related genes. They provide
a link between our thoughts and our bodies, between biology and
psychology, and are an important piece of the puzzle of how psychological states can affect our bodies, and vice versa. They also explain
how psychotherapy, prayer, worship, and social rituals can have positive effects on our physical wellbeing. They offer a pathway by which
we can influence physical health by immersing ourselves in behavioral
states that promote health, and avoiding behavioral states that can
hurt us.

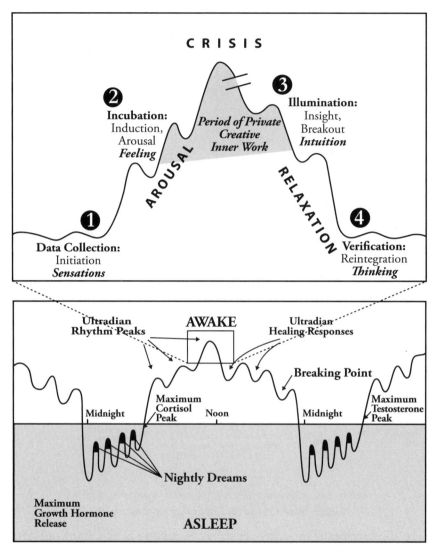

Ultradian rhythms. Gray area in top box shows peak of cycle;
grey area in bottom box shows sleep[15]

Therapeutic experience suggests that behavioral states are usually linked to daily *(circadian)* rhythms, or to periodic, several-times-per-day *(ultradian)* rhythms. Circadian rhythms follow the 24-hour clock, subject to the modifications of circumstance. Ultradian rhythms are briefer rhythms coordinated within circadian rhythms. They last about 90 to 120 minutes, and correlate with measures

84

of our energy level such as blood glucose, metabolic rate, hormone release, and insulin production.[16] Peak activity of the left and right hemispheres of our brains also alternate according to 90- to 120-minute ultradian rhythms,[17] and when we go to sleep, REM or dreaming periods follow a similar schedule.[18] Ninety to 120 minutes is also the average time between gene expression, and the synthesis of the proteins required by the body to convey information between cells, provide energy, create the scaffolding of cells, and many of the body's other functions.

Rossi notes that they correspond with a need for relaxation after periods of intense creative work,[19] and recommends an ultradian rest period in the afternoon, if that is when behavioral problems recur.[20] By this late in the day, he believes that many people, after ignoring the peaks and valleys of their ultradian cycle for many hours, have "an accumulated ultradian deficit and stress syndrome expressed with these common complaints:

"'I'm exhausted by mid-afternoon."

"'I get stressed, tense, and irritable toward the end of the workday."

"'I need a drink after work."

"'My addiction gets worse later in the day when I have to *have* something."

"'I get sleepy in the afternoon."

"'The worst time is when I have to go home after school and I'm too tired to do homework."

"'Just before dinner everybody is irritable and that's when arguments start."

"Many of these acute and chronic problems can be ameliorated by taking one or two ultradian breaks earlier in the day or taking a nap after lunch,"[21] he advises, especially in cases where people might already have skipped several ultradian rest periods during which their bodies were clearly instructing them to slow down. Noticing our need for ultradian rest periods after times of intense creative output can allow us to pace our days in order to avoid behavior-dependent genetic conflicts.

Secretion of hormones like ACTH and cortisol, which are released on the usual 90- to 120-minute ultradian cycle, peak just before wakefulness. Most researchers do not believe that these fluctuations are under our conscious control. Yet many people are able to decide, before they go to bed, exactly what time they will wake up.

Like many travelers, I set my internal alarm clock when I'm on a trip. I decide when I want to wake up each morning just before dropping off to sleep. When I wake up and look at the clock, I'm usually within a few minutes of my target time. If I want to then sleep some more, I'll say to myself, "I'll wake up fully in twenty minutes"—and usually I will. The certainty that this biological function will work reliably is so engrained in my awareness that I long ago stopped carrying a travel alarm clock in my kit. This experience, common to a great many travelers,[22] suggests that *intentionality conditions aspects of our behavioral state-related genetic activity long thought to be outside of our conscious control.*

Experience-Dependent Genes

Experience-dependent genes are genes that are activated by learning and novelty. This class of genes generates the protein synthesis required to instruct stem cells to differentiate in order to replace injured or damaged cells in the tissues of our muscles and organs—the foundation of growth and healing. It also stimulates stem cells into forming new neurons in the brain—not just in the young, but at any age. Stimulated by novel activities and learning, these new neurons form new synaptic connections within the brain. The experiences we are having each moment are actually changing the structure of our brains.

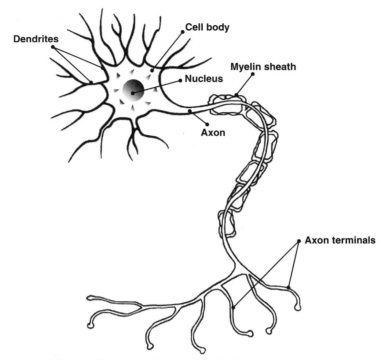

Nerve cell axon terminals transmit signals to other neurons

Experiences build neuronal pathways. When you first learned the correct way to swing a tennis racket, it took great concentration to remember your instruction each time. Then, at some point, the neural synapses that coded your swing were so well developed that you no longer had to concentrate. The nerve connections associated with swinging the racket correctly were copious enough to permit you to perform this feat without conscious attention. Step onto a tennis court, and your body immediately knows how to swing a racket. You could move on to another learning experience.

Contrary to the popular notion that "you can't teach an old dog new tricks," our brains keep adding new neural links throughout our lives, as long as they are stimulated to do so.[23] This process is called *neurogenesis*. Learning experiences and other *highly attentive states of awareness* switch on the expression of genes that stimulate the formation of new neurons. While most of our organs stop growing in our late teens, our brains—with the ongoing stimulation of new behavior, discovery, physical exercise, novel environments, and fresh new memories—are a teeming mass of creation our whole lives.

One of those creations may be health where before there was disease; "Many of the so-called miracles of healing via spiritual practices and therapeutic hypnosis (Barber, 1990), probably occur via this type of activity-dependent gene expression in stem cells through out the brain and body."[24] Rossi declares that, "fascination during novel and numinous life experiences plays a fundamental role in focusing our attention and engaging activity-dependent gene expression neurogenesis, and healing in general."[25]

The process of turning short-term memories into long-term memories is key to neurogenesis. Short-term memory utilizes only *existing* pathways of molecular communication between nerve cells. Long-term memory, on the other hand, is processed by the portion of our brains known as the hippocampus. As it codes for long-term memory, the hippocampus stimulates experience-related gene expression, especially involving a gene known as zif-268, which leads

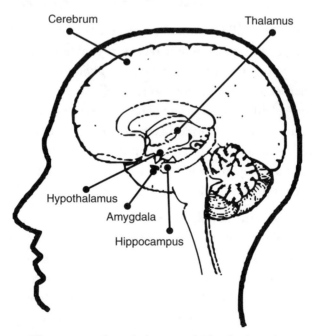

Hippocampus, hypothalamus and other brain regions

to the growth of new synapses and *new neural pathways* in the brain. London cabbies, who have to navigate a crazy quilt of medieval streets each day, tend to have larger hippocampuses.[26] So do symphony violinists. Conversely, long-term stress, with its exposure to cortisol, has been shown to damage the hippocampus, inhibiting memory and learning.

New techniques that allow us to produce images of brain activity, like Positron Emission Tomography (PET) scanners, and functional Magnetic Resonance Imaging (fMRI) machines, are giving us a new understanding of how the brain works. As subjects think certain thoughts, practice certain behaviors, or harbor certain emotions, researchers can determine which areas of their brains are firing. Emotions such as fear and anger are associated with different patterns of brain wave predominance.[27] We are now also starting to be able to associate changes in the brain with changes in genetic state; as certain genes are activated, certain areas of the brain show increased activity. This genetic activation produces many other changes in our cells.

Candace Pert, Ph.D., author of *Molecules of Emotion,* calls this loop between environment and cells the *psychosomatic network*. Through the psychosomatic network, thoughts and emotions are transformed into physiological effects. In the other direction loop, physiological experiences can translate into mental and emotional states.

Burying Trauma in Muscles

"I am looking at and seeing the violent crash. I see my father's body across the railroad tracks. I feel the shock and horror in my mother's body and consciousness as she witnessed his death."[28]

These words were spoken by a woman receiving a head massage. Her face "contorted in agony" as she suddenly, vividly, and spontaneously recalled the death of her father when she was a child.

For decades, massage therapists have recounted such stories of spontaneous awakening of memory when tissues are stimulated. They can scarcely fail to notice the link between body and mind; when manipulating muscle and connective tissue, emotional release of buried traumatic memories sometimes takes place. Science is now catching up and starting to describe some of the mechanics of this phenomenon. It has also become apparent to many psychotherapists that verbal processing of trauma, without physical release, provides only partial relief. There are many accounts of buried emotional traumas being released by means of bodywork; there are also reciprocal accounts of physical afflictions clearing up once a psychological shift occurred.

Stanislav Grof, M.D., Ph.D., who coined the term "spiritual emergency" to describe a dramatic spiritual breakthrough manifesting as a psychotic episode, said that spiritual emergencies might be accompanied by involuntary twitching and other spontaneous body movements. Harvey Jackins, who founded one of the most widely used forms of peer therapy, known as Re-Evaluation Counseling or C0-Counseling, believes, on the basis of thousands of clinical observations, that genuine psychological shift is *always* accompanied by shuddering, moaning, twitching, tears, sweating, or some other physical sign of discharging emotions. A 2004 meta-analysis of thirty-seven studies of massage therapy published in *Psychological Bulletin* showed their effectiveness for the relief of anxiety and depression, with "benefits similar in magnitude to those of psychotherapy" alone.[29]

Using both psychological counseling and physical manipulation in tandem may provide the most effective treatment for many people. Candace Pert reminds us that, "the body is the unconscious mind." Psychotherapy that ignores the body may make arduous and ineffective therapeutic detours around the direct route.

The link between areas of the body and traumatic emotional states is illustrated by many phenomena. One is the recent discovery that removing "worry wrinkles" may remove the underlying worry too. This effect was stumbled upon by plastic surgeons giving patients cosmetic injections of Botox. Botox, a therapeutic variant of the protein present in botulism toxin, paralyzes muscles into which it is injected. When injected into the facial muscles of patients with deeply lined skin, it paralyzes the muscles, and the skin smoothes out for a few months.

What cosmetic surgeons began noticing, however, was that in some of their Botox patients who were depressed, the depression lifted after the injection. According to one report:

> Kathleen Delano had suffered from depression for years. Having tried psychotherapy and antidepressant drugs in vain, she resigned herself to a life of suffering.
>
> Then she tried Botox, the drug that a few years ago became the rage for smoothing facial wrinkles.
>
> In 2004, her physician injected five doses of the toxin into the muscles between Delano's eyebrows.... Eight weeks later...her depression had lifted.[30]

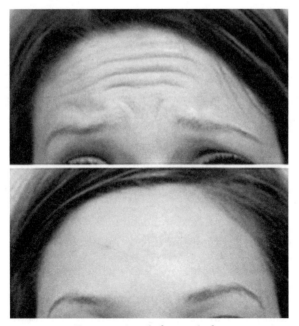

Botox patient before and after

The alteration of the physical structure seemed to catalyze an alteration of the mental state. A happier "look"—even one artificially induced—produced a happier experience. While we know that changing the mind changes the body, it also seems that the reverse is true: Changing the body changes the mind.

During a therapy session with "Celeste," a young woman with arthritis chronicled in Ernest Rossi's *The Psychobiology of Gene Expression,* Celeste goes from very limited mobility in her hands, to making a fist, to being able to stretch her fingers wide. She's delighted by the changes in her body that occur during the hour-long session with Rossi. And while he acts as a therapist, he also speculates which genes are expressing:

> **Celeste:** My right hand is doing some stuff.
>
> **Rossi:** Your right hand is doing some stuff?
>
> **Celeste:** Yeah. My left one feels like lead, but my right one.... I don't know, I think it started shaking a little bit, or something.

Rossi's comments of this exchange: "Evidence of psychobiological arousal and behavioral state-related gene expression.... The therapist wonders how to engage the psychogenomic dynamics of immunological variables such as interleukin-1, 2, and 1B associated with Cox-2 that has been implicated in rheumatoid arthritis that is Celeste's presenting condition."

> **Rossi:** [Celeste's hand...surprisingly forms a fist]. Oh my goodness? Something new seems to be happening?
>
> **Celeste:** Yeah!...
>
> **Rossi:** Wow! Yes, something new is beginning to happen! [Celeste now extends her fingers up into the air]....
>
> **Celeste:** I sure don't know what this is [laughing]....

Rossi's comments: "Illumination and activity-dependent gene expression. Celeste experiences playful activity-dependent exercise as a creative breakout of her typically restrained hand and finger movements associated with her rheumatoid arthritis. Future research will be needed to determine if...the CREB genes associated with new memory and learning—as well as the ODC and BDNF genes associated with neurogenesis and physical growth are actually being engaged...."

At the end of the session, Celeste is stretching her arms and hands in delight at her newfound mobility. Rossi closes by hoping that, "the experiential theater of demonstration therapy will be sufficiently numinous to activate zif-268 gene expression in her REM dreaming tonight..."[31] which will help cement the changes in her body.

Recent studies are allowing us to understand the role of genetics in the psychosomatic network. For instance, the hypothalamus is a structure in the brain that transforms the activity of the frontal lobes into hormonal messenger molecules. These communicate with the endocrine glands, which affect other systems, including the immune system, digestive system, and muscular-skeletal system. Portions of the hypothalamus synthesize a hormone known as CRH (corticotrophin-releasing hormone), which stimulates the production of a dozen other messenger molecules that influence stress and relaxation.

The gene that initiates the production of CRH is located on chromosome 8, a chromosome so vital to our function that is has scarcely changed for millennia: "CRH synthesizing and secreting neurons are found [in] their highest densities...in the prefrontal, insular, and cingulate areas [of the brain], where they mediate cognitive and behavioral processing.... Secondary messengers within the cell *convey the extracellular signals from the environment (including psychosocial cues) to the nucleus of the cell, where they initiate gene expression.*"[32] Our bodies have an exquisite ability to turn external cues into the signals required for optimal internal responses.

Eric Kandel, who received the Nobel Prize in Medicine in 2000, says that "Changes in gene expression...alter the strength of synaptic connections and structural changes that alter the anatomical pattern of interconnections between nerve cells of the brain."[33] In one experiment, Kandel discovered that when new memories are established, the number of synaptic connections in the sensory neurons stimulated jumped to around 2,600, a *doubling* of its previous count of 1,300. Unless the initial experience was reinforced, however, the number of connections dropped back to 1,300 within three weeks. If we reinforce our novel experiences by repetition, we strengthen the neural net to support them; if we do not, our newfound neural circuitry quickly decays—not over years, but *in less than a month.* This means that new thoughts, actions, and habits must be continuously updated in order to take root.[34] Like a song on the radio that becomes ingrained in the cultural collective, many people must hear it many times. But if it isn't heard for a while, the memory of even a Top Forty tune begins to fade: who today remembers "It's a Long Way to Tipperary?" or "Way Down Upon the Swanee River?"—both top of the charts a century ago.

Insight changes the brain as well. In an article entitled *The Neuroscience of Leadership,* David Rock and Jeffrey Schwartz, Ph.D., report on studies that use, "MRI and EEG technologies to study moments of insight. One study found sudden bursts of high-frequency 40 Hz oscillations (gamma waves) in the brain appearing just prior to moments of insight. This oscillation is conducive to *creating links across many parts of the brain.*" This is the part of the brain that "is involved in perceiving and processing music, spatial, and structural relations (such as those in a building or painting), and other complex aspects of the environment. The findings suggest that at a moment of insight, a complex set of new connections is being created."[35]

Gene expression in long-term memory encoding also has an ultradian rhythm of between 90 and 120 minutes, and the number of new synaptic connections between neurons can double in as little as an hour once the experience-dependent genes are activated.[36] One of the ways in which memories are encoded is when we replay a scene in our minds. Memory is not static, and as we *combine old memories with present situations,* we stimulate neurogenesis.

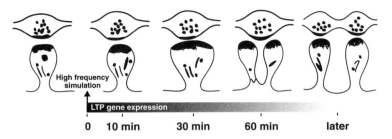

Within an hour of gene expression in response to environmental stimulation, one synaptic connection has become two[37]

The bad news is this: if you don't use these connections, you lose them. "Should synaptic stimuli fall below a certain threshold indicative of disuse, it would be appropriate for synaptic connections to be broken..." asserts one researcher.[38] Kandel's work also showed that if novel learning experiences were not reinforced, the base line of 1,300 synaptic connections could drop to 800.[39]

Novel experiences lead not only to the growth of new brain tissue; they are linked to psychological wellbeing, too. There is a link between clinical depression, and "a lack of new cell growth in the brain."[40] As a consequence, the hippocampi of the brains of depressed patients shrink in time by as much as 15%, as distress and social trauma result in environmental signals that inhibit the expression of experience-dependent genes.[41] Our hippocampi are involved in the recall of memories, perhaps while we sleep; such repeated replaying is central to the process of creating durable long-term memories,[42] with a beneficial effect on neurogenesis. Psychologist Martin Seligman, Ph.D., in his book *Authentic Happiness,* sums it up by saying "Neurons are wired to respond to novel events."[43]

Rebooting Memory for Healing

The fluctuating nature of memory, with synaptic connections being created and destroyed, means that memories may be strengthened or diffused. Psychotherapy seeks to bring painful memories back to the forefront of consciousness, and then shift them. A series of experiments studying rat brains suggests that if a painful

95

or fear-laden memory is triggered and then processed, its impact is diminished. Rossi, who has more than three decades of experience treating patients, summarizes these experiments as follows: "When the rat brain is infused with anisomycin (an inhibitor of protein synthesis) shortly after the reactivation of a long consolidated memory, the memory is extinguished. The same treatment of the brain with anisomycin but without reactivating the consolidated memory leaves the memory intact. This means that the gene expression and protein synthesis cycle is reactivated when important memories are recalled and replayed...."

"Most paradigms of psychotherapy," he continues, "involve a combination of the same two-step process.... (1) a reactivation of old traumatic memories, which is (2) immediately followed by some form of therapeutic intervention designed to heal the old hurt."[44] Each time we reboot an old memory, we may be unconsciously modifying it even as we think we are "just remembering." This phenomenon can help with the healing process, especially when combined with strong feelings; we can then cement an association between an old trauma and a new, positive meaning.

One study showed an increase in immune system function during a drumming and storytelling ritual. In this study, 111 healthy volunteers were exposed to an hour-long ritual, much like the kind of communal experience our ancient ancestors might have enjoyed around the campfire. The improvement in immune function after the drumming and storytelling ceremony was demonstrated by increased activity in helper T-cells. In addition, levels of the healthy hormone DHEA increased, while the stress hormone cortisol dropped.[45]

Many traditional healing rituals, such as shamanic journeys, faith healings, passion plays, the Catholic Mass, and exorcisms, involve the same mechanism of heightened emotional arousal, followed by a release. Rossi points out that psychotherapies old and new use this approach. Sigmund Freud first had patients free associate, then, in the second step, find an insight that reframed the meaning of the memories in a significant way. New therapies like Eye Movement

Desensitization and Reprocessing (EMDR) also reactivate painful old traumas, after which they are infused with positive images and feelings.[46]

Creating Your Own Designer Brain

Learning is demanding. Giving ourselves new challenges, like taking a college course in a subject completely beyond our existing fields of expertise, reaching out for new friendships, acquiring a new artistic ability, and learning a new sport, all stretch our consciousness. Yet this positive stress is part of the process by which we grow the capacities of our brains. Experiments with rats being taught new tasks found that those rats that were mildly stressed learned faster than those that were not.[47]

Other experiments demonstrate that *unpredictability* and *novelty* is a crucial aspect of learning. The association between a stimulus and a response—a known reward for a known action—takes learning only to the first step. After that, stimuli must be unpredictable in order to maintain responses, and continue engaging experience-dependent genes. If we *expect* a reinforcement for a certain response, the novelty value of that reinforcement quickly wears off, and learning stops. It is the unexpected that commands our attention, not the known.[48]

Rossi calls this the *novelty-numinosum-neurogenesis* effect, and characterizes it as a "core dynamic of psychobiology. It integrates experiences of mind (sensory-perceptual awareness of novelty with the arousal/motivational aspects of the numinosum) with biology (gene expression, protein synthesis, neurogenesis, and healing).... Activity-dependent creative experiences in the arts, cultural rituals, humanities, and sciences as well as the peak experiences of everyday life are all manifestations of the novelty-numinosum-neurogenesis effect. When reviewing awesome art or architecture, when moved by cinema, music, and dance, when enchanted by drama, fantasy, fairytale, myth, or poetry, we are experiencing mythopoetic transmissions of the numinosum...."[49] Every time you expose yourself to such learning experiences, you are taking an important step toward health and long-term mental acuity.

A fascinating recent series of experiments examined how we can fool ourselves into believing we saw or did something when, in fact, we did not—and those beliefs translate into neurochemicals. In one study, 148 young British college students were served in a bar. Everything about the bar was real: the bottles, the glasses, the napkins, the sights and smells.

Unbeknownst to the experimental group, there was one thing that was fake: the alcohol. Researchers had substituted the alcohol in the bottles of spirits, beer, and wine with tonic. The bartenders mixed the drinks as though they were serving the real McCoy, and *the subjects became tipsy,* acting in a manner similar to the control group, who were being served the real thing. Their bodies generated the neural signals and neurochemicals resulting in intoxicated behavior *simply because the students held the belief* that they were drinking real alcohol. "'When students were told the true nature of the experiment at the completion of the study, many were amazed that they had only received plain tonic, insisting that they had felt drunk at the time,' the researcher commented, concluding that, 'It showed that even thinking you've been drinking affects your behavior.'"[50]

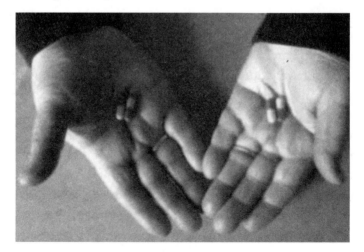

Which is the placebo?

98

In another study reported in *New Scientist,* fourteen healthy young men were given an injection in their jaws that produced pain, plus a pill that they were told "may or may not relieve pain." The brains of the subjects produced endorphins, the body's natural opiates, even in those subjects given a placebo. Their belief alone was sufficient to fire up their body's natural biochemical plant to produce pain-relieving substances.[51]

What's happening in our brains can override what's happening in our bodies. Under hypnosis, subjects can be induced to really "see" things that aren't there. They can be trained to look at common English words, and perceive them to be gibberish. In a provocative summary of this research published in the *New York Times,* Dr. Amir Raz said that the brain's internal beliefs and perceptions, "'overrode brain circuits devoted to reading and detecting conflict.' A number of other studies of brain imaging point to similar top-down brain mechanisms.... Top-down processes override sensory, or bottom-up information, said Dr. Stephen M. Kosslyn, a neuroscientist at Harvard. People think that sights, sounds, and touch from the outside world constitute reality. But the brain constructs what it perceives based on past experience, Kosslyn said."

Researchers have noticed that we can still read and make meaning out of words even if they are jumbled. Read the following three paragraphs quickly:

Olny srmat poelpe can raed tihs.

I cdnuolt blveiee taht I cluod aulaclty uesdnatnrd waht I was rdanieg. The phaonmneal pweor of the hmuan mnid, aoccdrnig to rscheearch at Cmabrigde Uinervtisy, maens taht it deosn't mttaer in waht oredr the ltteers in a wrod are, the olny iprmoatnt tihng is taht the frist and lsat ltteer be in the rghit pclae. The rset can be a taotl mses and you can sitll raed it wouthit a porbelm.

Tihs is bcuseae the huamn mnid deos not raed ervey lteter by istlef, but the wrod as a wlohe. Amzanig huh? And I awlyas tghuhot slpeling was ipmorantt! Now you can tlel tehm taht it inst! And wlihe yuo're at it, plseae dno't fegrot to tlel all yuor fneidrs taht tiher leivs wlil be mcuh bteter if tehy buy at lseat ten cepois of *The Ginee in Yuor Gnees.*

You probably found that you could scan and comprehend them almost as fast as if they had been unscrambled. There are also many visual tricks like the one below. An object or word can seem one thing, or seem another, depending on our perceptions.

Good or evil?

A team at Yale came to the startling realization that "the cortical map reflects our perceptions, not the physical body," adding that, "the brain is reflecting what we are feeling, even if that's not what really happened." Anna Roe, the chief researcher, said, "We think we know what's out there in the physical world, but it's all interpreted by our brains. Everything we sense is an illusion to a degree."[52]

One possible mechanism to explain the ability of our brains to override our senses is the surprising recent discovery that the bundles of nerve cells running from our brain to our *senses outnumber the ones running in the other direction* by a factor of ten to one! For every neural bundle running from our senses to our brain, there are roughly *ten* neural bundles running from our brain to that sensory organ. So there's a lot more bandwidth for signals going *from* the brain than there is for signals going *to* it. University of Oregon neuroscientist Michael Posner, Ph.D., says, "The idea that perceptions can be manipulated by expectations is fundamental to the study of cognition."[53]

Since we are building these neural pathways with every thought and feeling, we have an opportunity, by taking control of the quality of our thoughts and feelings, to build a neural network focused on the transmission of positive, healing, and joyful impulses. As we consciously cultivate these mental and emotional states, which then become ingrained in our neural network, we may indeed see, in time, the beautiful world we imagine. A wise fool exclaimed: "If I hadn't believed it, I wouldn't have seen it with my own eyes."

Designing Happiness

Far from who and what we are being determined by our genes, we are rewriting the expression of our genes in every second, by our choices of what to do, say, and think. The choices we make with our consciousness are being genetically encoded in our brain structure daily, reinforcing the pathways that correspond to experiences we have frequently, and reducing pathways we use infrequently. But more than the "use or lose it" axiom, activity-dependent genetic expression tells us that we can "experience it and create it" as we encode new pathways in our brains that correspond to a particular experience.

This research places primary responsibility for health and healing back in the hands of the individual. It makes us aware that it is not doctors, hospitals, acupuncturists, homeopaths, chiropractors, energy workers, or other health professionals who are responsible for our wellbeing. Health begins with us.

We must consider the implications of the fact that our emotional and mental environment, which we create as individuals, is one of the primary influences turning genes on and off in our cells. The ever-changing, synchronous flow of opportunities that this vision opens up for us are so vast and exciting that it will take medicine decades to grasp the full scope of these potentials. While it may require hundreds of scientific studies to chart the links between specific environmental influences and specific genetic activations, we as individuals can start to use this knowledge to affect our own health right now. The awareness that we each create our mental environment, and that the genetic effects of environmental changes begin to occur within seconds, provides us with an exciting new awareness of the degree of control we enjoy over our wellbeing.

It also counteracts the prevailing feeling of helplessness that many patients feel when enmeshed in the medical system. When we realize that we have conscious control over the biochemical environment in which we bathe our cells, we suddenly become acutely aware of which ingredients we are dropping in the stew. Like the cook mixing the stew, we can choose to put only tasty thoughts and feelings into our cells. We would no more put toxic thoughts into the stew of our consciousness than we would throw rat poison into the stew.

We likewise become aware of which emotions we harbor. We perceive emotions not simply as experiences that happen *to us,* but as aspects of our environmental cocktail that we can *cultivate* to bring beauty and nourishment to the garden of our health. Over the last few decades, research has sought to understand the mechanisms that underlie the health effects of positive consciousness. We know that altruism, optimism, prayer, meditation, spirituality, social connectedness, and energy medicine have positive effects on health and longevity. Now we're starting to understand that our consciousness conditions our genetic expression, moment by moment. This insight allows us to *use consciousness change as a medical intervention.* For instance, studies have shown meditation to have benefits that are similar to anti-depressant medications, regulating the serotonin and dopamine levels in our brains, as well as stimulating our immune systems.[54] Other research shows that we can alleviate chronic pain through meditation, and that our brain's response to pain is reduced by meditation.[55] Knowing that we can unlock a hugely beneficial internal pharmacopia of gene-altering, naturally occurring substances through consciousness—without any of the side effects of artificial drugs—gives us powerful leverage for wellbeing.

Robert Beck, D.Sc., was the first to realize, in the 1960s, that charts of the oscillations of the Earth's magnetic frequencies looked a lot like EEG readouts from humans.[56] He performed experiments with healers from various regions and religions, including Amazonian shamans, Hawaiian kahunas, Christian faith healers, Indian yogis, and

Buddhist lamas, and showed that—at the moment of healing—their brain wave frequencies were virtually identical.

The Earth resonates at an average frequency of 7.8 Hz,[57] while, "the dominant brainwave frequency of sensitives, such as shamans and healers, comes close to 7.83 Hz and may, at times, beat in phase with the Earth's signal, thereby causing harmonic resonance."[58] Before and after the healing moment, the EEG scans differed from each other in typical ways. But during the healing state, they all shifted to reveal similar dominant frequencies. It doesn't seem to matter what the particular belief system is that puts the healer in that state; they call on a range of healing entities from Kwan Yin to Jesus Christ to White Buffalo Calf Woman. Whatever belief system shifts the healer into the healing state, once there, their brain frequencies show similar characteristics.[59] They appear to be tapping into a universal frequency that is implicated in healing, and is effective regardless of the belief structure of the healer.[60] Long before we had medicines, human societies must have discovered ways to catalyze healing that were dependent on electromagnetic alignment with the Earth itself.

An excellent and informative survey by researcher Leane Roffey, Ph.D., of over 150 studies of "healing energies," some of which measured the electromagnetic energy emanating from the hands of healers, found that over half of them demonstrate a significant effect.[61] Rossi believes that the ubiquity of healing practices across time occurs because, "Healers in many different cultures developed and utilized many different worldviews and belief systems to deal with essentially the same question: 'How can we use human consciousness, psychological experiencing, and our perception of free will to communicate with our bodies in ways that facilitate healing and wellbeing?'"[62]

When we grasp the enormous opportunities for altering our genetic expression offered by changes of consciousness, we reclaim responsibility for our wellbeing and give ourselves options. We no longer have an imperfect medical system as our only hope for healing. There is nothing wrong with the people and the professionals in the conventional medical system, and there is nothing wrong with most of the techniques of allopathic medicine. They are an important piece of the treatment puzzle—just ask anybody who's had to go to the emergency room with a broken leg.

What is frightfully wrong is to presume that conventional allopathic medicine is *the whole* of medicine. This is like the drum thinking that it's the whole orchestra: while it might be a fantastic drum, providing exactly the right drama and exactly the right effects during its assigned parts in the symphony, when it starts to sound continuously, the result is a cacophony that masks subtler sounds.

Just as you wouldn't try to treat appendicitis with the laying on of hands, there are a great many medical complaints for which the first response might be a consciousness-based intervention—even before making a visit to your physician to rule out a medically treatable organic cause. This, as it turns out, is sound science as well as sound medicine. Research demonstrates that *only 16%* of patients visiting medical clinics have an identifiable organic ailment.[63] Later in this book we will also look at which specific classes of conditions are best treated with allopathic medicine, and which are best treated with energy medicine. As we look at the tip of the iceberg of the state of medicine in later chapters, we will see that putting the drum back in its perfect and appropriate place in the symphony is an essential step toward taking back responsibility for our individual wellbeing—and releasing the illusion that responsibility ever lay anywhere else.

4

The Body Piezoelectric

I'm deeply sure that we will never be able to understand the essence of life, if we restrict ourselves to the molecular level.... A surprising subtlety of biological reactions is stipulated by the mobility of electrons and can be explained only from the position of quantum mechanics.

—Albert Szent-Gyorgyi, Nobel Laureate, 1968

Electricity is at the foundation of medicine, ancient and modern. Electrical pulses from heart pacemakers regularize the heartbeats of tens of thousands of cardiac patients. Electromagnetic devices like MRIs, EEGs and EKGs allow doctors to scan the insides of patients' bodies without resorting to risky and invasive surgery.

The effects of electrical shocks were observed by the Greeks and Romans more than two thousand years ago. In his dialog *Meno,* set down around 400 B.C.E., Plato described stupefying electric fish, the torpedo fish, that lived in the Mediterranean Sea. His near-contemporary Aristotle indicated that the torpedo fish would narcotize its prey. In 100 C.E., the Roman writer Pliny also commented that the torpedo fish, while not sluggish itself, would induce torpor in other fish, and described "how to extract the medicinal ingredient of the torpedo fish into oils and ointments used for various ailments."[1]

Piezoelectricity is one of the most fascinating ways of generating electricity, and one that is essential to the understanding of the mechanisms of healing. Piezoelectricity is generated by mechanical means. When pressure is applied to certain structures, they polarize into positive and negative electrical poles, and generate electricity.

Cardiac pacemaker:
a common application of an electromagnetic field in medicine

A common application of piezoelectricity is the lighter in your home barbecue, or a modern gas grill. When you click the "light" button on your stove, you hear a clicking sound. That sound is a ceramic element being struck. The ceramic material is piezoelectric; in response to mechanical stress, it produces an electric spark that ignites the gas used in the burners.

The first documented use of piezoelectricity was by the Ute Indians, who lived in what is now the American state of Colorado. They created hollow rattles made of buffalo hide, into which they inserted quartz crystals. When the rattle was shaken, the quartz crystals struck each other, creating a mechanical stress that generated a piezoelectric discharge in the form of light. The light shone through the translucent skin of the rattles. During sacred ceremonies, thousands of years ago, the rattles would be shaken, and would glow in the darkness of the Colorado night, calling the spirits into the sacred circle.

Ute quartz rattle

Marie Curie was the first woman in France to receive a doctoral degree. She and her husband Pierre Curie discovered the medicinal effects of radiation, and were the first to coin the term "radioactivity." Together with Jacques, his brother, Pierre Curie first demonstrated the effects of piezoelectricity in 1880 by showing that many different types of crystals generate minute currents of electricity when subjected to mechanical stress. Marie's medical research leading to the development of practical X-ray machines was facilitated by the development of sensitive piezoelectric measurement devices by her husband. Pierre and Marie Curie received the Nobel Prize for physics in 1903 for their work. Marie also later received a second Nobel Prize, in 1911, becoming the first person to win or share two Nobel prizes.

The First World War saw the first large-scale application of piezoelectricity, with the development of sonar. Developed to track submarines underwater, sonar devices employ the piezoelectric phenomenon in the form of quartz crystals sandwiched between two steel plates. In this case the reverse of the phenomenon is utilized.

Applying electricity to the crystal produces a sudden motion that results a sharp chirp being sent out underwater. By measuring the length of time it takes for the echo of that chirp to bounce off the hull of another vessel and return to its point of origin, the sonar operator can determine the distance of the other vessel. As the distance changes over time, the sonar operator can also determine the other vessel's speed.

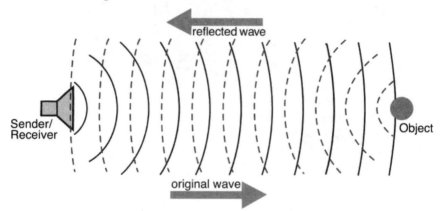

*A sonar signal is emitted, bounces off an object,
and returns to the originating source*

The success of sonar a century ago prompted intense interest in, and development of, piezoelectric devices. Among them were phonograph needles, microphones, and television remote controls. The piezoelectric principle is used today in thousands of advanced applications, including echo-locating units in some new cars that sense the distance between the front bumper and an approaching object, and can sound a proximity warning alarm.

So what are gas grills, car electronics, and Indian rattles doing in a book about the link between consciousness and your cells? The reason is simple: the human body is also a piezoelectric generator, and some of the structures in your anatomy have, as one of their primary purposes, the conduction of piezoelectricity from one part of your body to another. Energy is the currency in which all transactions in nature are given or received.

When you sign up for a massage and lie on the table having your muscles stroked, you are having a direct experience of a piezoelectric effect on the human body. Any kind of mechanical stimulation of the body creates piezoelectricity. This includes not-so-pleasant sources of piezoelectric stimulation, such as banging your shin against a table while navigating a dark room. Either of these experiences creates a piezoelectric charge in the cells surrounding the point of contact, and a piezoelectric current that travels along the most conductive channel available within the body.

Many kinds of tissue in our bodies have piezoelectric properties, including bone,[2] actin, dentin, tendon, and our tracheal and intestinal linings, as well as the nucleic acids of our individual cells.[3] Information is continuously flowing through your body in the form of electromagnetic currents. While there are many varieties of electromagnetic activity generated by the human body, and that may affect the body from outside, piezoelectric induction is a primary method by which electricity is generated in the human body.

Electrical Medicine in Western Science

The link between electrical currents and healing has been recognized for centuries. The term "electricity" was coined by William Gilbert in 1600 in his book *De Magnete*. A few decades later, electricity was being artificially generated, stored, and transmitted. Stephen Gray, in the early 1700s, showed that some materials were better conductors of electricity than others. Another Stephen, Stephen Hales, applied this insight to physiology, speculating that nerves might conduct electricity within the body. It was not long before adventurous researchers began to apply electricity to medicine. In 1753, Johann Schaeffer published *Electrical Medicine,* the first book on the subject. And Luigi Galvani demonstrated in 1786 that static electricity generated outside the body could be conducted through the nervous system to produce the contraction of a muscle within the body.

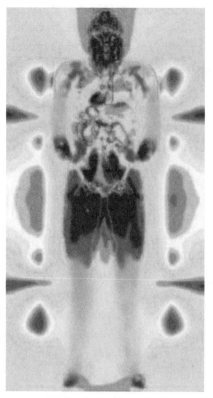

Human Magnetic Field

In the 1830s, Carlo Mateucci demonstrated that injured tissues generate an electrical current, and in 1868, Jules Bernstein first described "bioenergy," in which positively charged ions generate electricity as they move across cell membranes. He discovered that negatively charged chloride ions cling to the inside of a cell membrane, while positively charged sodium ions cling to the outside of the membrane. When a nerve receives an impulse from an adjoining synapse, the polarization reverses momentarily. This change of polarization ripples down the axon like an electrical current, transmitting energy from one end of the neuron to the other. However, the chemical nature of this exchange led scientists toward chemical explanations for the body's signaling systems, and electromagnetic signals that occurred outside the well-understood mechanisms of the nervous system began to receive less experimental scrutiny.

One of the nineteenth century's most influential researchers, Emille Du Bois-Reymond, constructed a device to demonstrate the effect of electricity, conducted through nerves, on living tissue. A species of fish known to emit electrical impulses was wired to the nerves of a frog's leg. When the fish generated a current, the frog's leg twitched, pushing a lever, which rang a bell—history's first example of a biotechnology machine.

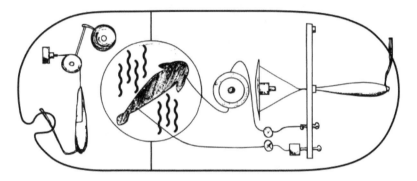

Nineteenth-century demonstration of the
conduction of electricity in the nervous system

The application of electricity to medicine continued in the twentieth century. In 1902, another Frenchman named Leduc reported that he could narcotize animals with a current of 35 volts AC at a frequency of 100 cycles per second. In 1924, Willem Eindhoven, a Dutch physician, won the Nobel prize for his discovery that the heart generated its own electromagnetic field. At that time, measurement of the heart's field required the use of the most sensitive instruments available; today scientists are able to detect electrical and magnetic fields millions of times fainter.

In 1929, Hans Berger, using progressively more sensitive galvanometers, was able to detect and describe the brain's electrical field, which is much fainter than the heart's. His work contributed to the development of the electroencephalogram (EEG), which maps the electric field of the brain. Later, the magnetic fields of the heart and brain were mapped by magnetocardiograms and magnetoencephalograms. The heart's magnetic field is about 5,000 times stronger than that of the brain.[4]

Electromagnetic scanning allows us to see inside the body without invasive procedures

In the 1960s, Robert Becker, M.D., demonstrated improved regeneration of limbs in salamanders with the stimulation of an electrical current. This work was later applied to humans, where it was shown to reduce the time it took for bone fractures to heal. Dr. Becker's work helped to establish *the piezoelectric properties of connective tissue,* which *encases and joins all the other structures of our body.* In the 1970s, Nobel Laureate Albert Szent-Gyorgi reawakened interest in the subject. He reminded scientists that: "Molecules do not have to touch each other to interact. Energy can flow through...the electromagnetic field."[5] He was also "impressed by the subtlety and speed of biological reactions, [and] proposed that proteins may be semiconductors."[6]

Certain strains of bacteria orient themselves to the Earth's magnetic field. They have been shown to contain microcrystals of magnetite, a black mineral form of iron oxide. Particles of magnetite are the smallest magnets occurring in nature. Small crystals of this magnetic substance are present in the brains of certain animals that require the ability to orient to the earth's magnetic field, such as homing pigeons, bees, and migratory fish.[7]

Researcher J. L. Kirshvink and his colleagues discovered magnetite in human brain tissue cells in 1992.[8] It occurs in linear chains of up to eighty crystals, often attached to a membrane.

Chain of magnetite crystals

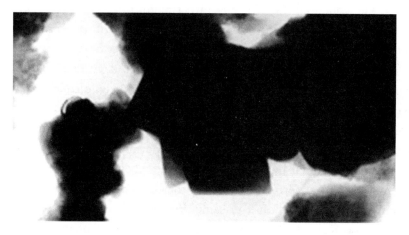

Magnetite crystals in human hippocampus

Magnetite microcrystals have been linked to our ultradian and circadian rhythms. We demonstrate this effect every time we fly a great distance and experience jet lag, as our body's circadian rhythms adapt to a new diurnal pattern, and a reorientation of our personal electromagnetic field (or EMF) to the patterns of the Earth's electromagnetic field that are unique to our new location. "In samples of human brain tissue cultured in labs, minute amounts of electromagnetic energy can affect brain tissue production of norepinephrine."[9] Norepinephrine is the neurotransmitter used in most of the sympathetic nervous system, and this research is being applied to the treatment of many neurological disorders.

When water is exposed to magnetic fields, then examined using infrared spectroscopy, it demonstrates reduced hydrogen bonding and other minute changes in its molecular structure. In a seminal series of experiments by Bernard Grad, Ph.D., of McGill University, these same molecular changes were demonstrated in samples of water on which healers had performed "the laying on of hands." Grad performed scores of experiments on seed germination and plant growth over the course of many years. He found that when seeds were watered with the water that had been held by the healer, they grew significantly faster and larger than those that had not.[10]

He wondered if the reverse effect might not also be true. So he compared the growth of control groups of plants with the growth of plants given water that had been held by patients being treated for psychotic depression. Sure enough, plants irrigated with that batch of water demonstrated very little germination, and significantly slower growth of those seeds that did sprout.[11] Grad later confirmed similar effects of the "laying on of hands" in the healing of cancers and wounds in laboratory animals.[12] Grad's experiments were later replicated and extended by other researchers.[13] They provide a foretaste of more recent experiments showing that the fields generated by the hands of healers exhibit the same frequencies as electromagnetic healing devices. Ciba-Geigy, a giant multinational pharmaceutical company, patented a process to create genetic changes in fish eggs—using only electrical effects. Using their process, "They were able to grow trout having distinctive hooked jaws that had been extinct for 150 years."[14] Both mechanical generators of electromagnetic fields, and human ones, can produce fields that shift genes.

Pulsed Electromagnetic Field Therapy

One of the most recent therapeutic uses for electromagnetism in healing is the use of pulsed magnetic stimulation (PMS). PMS machines deliver timed magnetic pulses at calibrated strengths to parts of the body, and are being tried for a wide variety of ailments, including the treatment of depression and inflammation. For this reason, it has been used to treat Alzheimer's disease, epilepsy, and Parkinson's disease.[15] Magnetic fields also appear to enhance the ability of diseased or damaged cells to utilize oxygen, thus speeding their recovery. The increased oxygen absorption may be linked to another effect of PMS stimulation, which is that patients report decreased pain almost as soon as the field is turned on.

Magnetic field therapy affects all the cells within its range. It travels through bone, beds, and plaster casts. Magnetic fields seem to work by affecting the electrical charge of cells. A normal cell has an electrical potential of about 90 millivolts. An inflamed cell has a potential of about 120 millivolts, and a cell in a state of degeneration may drop to 30 millivolts. By entraining the electrical fields of the cells within its range to the magnetic pulses emitted by the PMS machine, cells can be brought back into a healthy range. Not only are electrical changes observed in individual cells, they are observed in collections of cells, in organs, and whole systems, when healing is occurring. It appears that an electromagnetic field "intrinsically interwoven into the fabric of the system" assists in the signaling process of large numbers of cells in a coordinated healing process.[16]

Diseased cells show differences beyond their electrical charge. The biological software of invasive cancers, for instance, begins to produce an increasing number of glitches; they show an increasing degree of mutation in the DNA sequence. In the words of a cancer research team at the University of Illinois, Chicago: "there is a consensus in the field that increasing genomic deregulation also appears to be paradoxically associated with increasing malignancy among the most invasive and metastatic tumor cells."[17] The genetic profile of cancer cells, as well as their electrical charge, begins to deviate more and more from that of healthy cells.

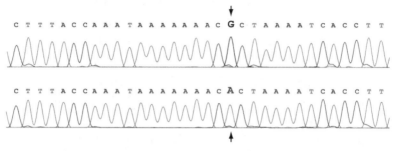

A genetic mutation in a sequence is indicated by arrows

Interestingly enough, in the case of certain types of brain cancer, epigenetic deviations are responsible for more tumors than DNA mutations. When the chemical profile of the proteins surrounding the gene does not replicate properly, tumor-fighting genes do not have the opportunity for expression. Healthy protein formation is as important as healthy gene formation, and points again to the importance of epigenetic control of the DNA replication process.[18]

One study reported on the results from sixteen medical centers using Pulsed Magnetic Stimulation to treat depression in 300 patients who had been unresponsive to SSRI drugs such as Prozac and Zoloft. Patients sit in a special chair for forty-five minutes per session, while magnetic pulses are directed at the parts of their brains linked to depression. Some 45% of the patients in the study experienced relief from depression using this method.[19]

Magnets have shown themselves effective in many kinds of therapy. But until the discovery of magnetite in the human brain, there were few credible explanations of why magnetic therapy "produced improvements in a wide range of medical conditions, including tendonitis, blood circulation, diabetic neuropathy, bone cysts, hypertension, optic nerve atrophy, facial paralysis, and fracture healing."[20]

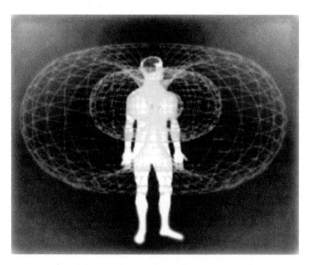

Electromagnetic field of human heart[21]

The Earth's predominant resonant frequency (the Schumann Resonance) is 7 to ten cycles per second (Hertz), with an average reading of 7.8 Hertz. This frequency is also common in the EEG readings of humans and many animals. Exposure to certain frequencies can trigger positive or negative stress, and effects on our immune systems.[22] Experiments by Robert Becker showed that electromagnetic pulses of certain frequencies alter the production of cortisol in the adrenal glands. "When people are removed from the normal 7–10 Hz background fields by placing them in specially shielded rooms, their EEGs, mood, and diurnal (day-night) neurochemistry change. The thyroid, pancreas, and adrenal glands are all affected by these EMFs."[23]

Other experiments have shown that when human beings are removed from these shielded rooms, their normal circadian and ultradian rhythms can be restored by exposing them to 10 Hertz fields.[25] Human DNA has a frequency of 54 to 78 GigaHertz, (GHz, or billions of cycles per second). Plant DNA has a frequency of 42 GHz, and animal DNA, 47 GHz. The frequency of human DNA is now being used by some clinical electrostimulation devices, and is reported to have great success in relieving pain in cases that have been unresponsive to conventional drug therapy. Electrostimulation devices that use the frequency of human DNA have just begun to see serious clinical scrutiny and application, and may come to play an important role in medicine in the future.[26]

Electrical Medicine in Direct Observation

In the 1950s, a German scientist named Reinhard Voll began to take careful measurements of the electrical charge at different points on the surface of the skin. He discovered that certain spots had a lower electrical charge than others. Voll's instruments discovered that these points of electrical difference corresponded with amazing fidelity to the acupuncture meridians described in ancient Oriental medicine. It is remarkable that the ancient practitioners of a five thousand-year-old healing tradition like acupuncture had the sensitivity to derive, *purely by observation,* a map of the energy pathways in the body similar to one demonstrated by modern scientific instrumentation.[27]

Ming dynasty meridian chart

120

Studies show that points on the meridians have much lower electrical resistance (averaging 10,000 ohms at the center of a point) when compared to the surrounding skin (which averages a much higher 3,000,000 ohms).[28] When the points are stimulated with a low-frequency current, the body responds by producing endorphins and cortisol. When they are stimulated with a high-frequency current, the body produces serotonin and norepinephrine. When the surrounding skin receives the same current, these neurochemicals are not produced.[29] Other studies compared the stimulation of acupoints with stimulation of areas close by that were not acupoints.[30] One of these studies showed that stimulation of the correct acupoints in people under stress resulted in lowered heart rate and a reduction in anxiety and pain. Another showed that depressed patients were helped by meridian stimulation, with an amazing 64% reporting *complete remission* of their depression.[31]

This indicates that something beyond the placebo effect is at work when meridians are stimulated. Other clinical experiments bear this out,[32] as well as many accounts by amazed Western physicians. Isadore Rosenfeld, M.D., writing in *Parade* magazine, recounts the following story: "In 1978, I was invited to China to witness an open heart procedure on a young woman. She remained wide awake and smiling throughout the operation even though the only anesthesia administered was an acupuncture needle placed in her ear."[33] Studies with a new generation of fMRI machines are illuminating some of the changes in the brain that result from acupuncture, especially the role of the meridians in reducing pain and producing anesthetic effects.[34]

Voll's discoveries of the electromagnetic properties of meridians led to some powerful clinical applications, in the form of a diagnostic technique called Electrodermal Screening or EDS. EDS or bioresonance machines use sensors that measure the electrical potential at different points on a patient's skin; they are prevalent in Europe, though still rare in the United States due to disputes about their efficacy.

Voll, and those who developed later applications of his technology, discovered that patients who were exposed to an allergen, toxin, or pathogen would show a change in the electrical fields of their skin. Berney Williams, Ph.D., of Holos University, an institution on the leading edge of energy medicine applications, has done extensive research on the electrical characteristics of the meridians. He believes that the "epidermis and dermis of the skin provides channels of heightened electron conductivity, which are experimentally measured as electrical conductance at acupuncture points on the surface of epidermis. Such channels can be theoretically present within the mass of connective tissue, which...might be associated with the transport of electron-excited states through molecular protein complexes."[35] EDS practitioners typically have a patient hold a vial of some substance to which they may be allergic. Readings from the machine show whether or not an allergic reaction is registered. By having patients pick up and put down dozens of vials, containing many common allergens, a profile can be developed of which substances may be triggering a harmful reaction.[36]

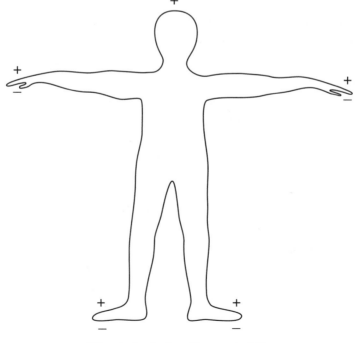

Electrical polarity of human body[37]

An interesting characteristic of EDS devices, and several allied technologies, is that they can detect *future* predispositions to certain diseases *before* they become manifest through physical symptoms. This predictive capability of energy readings was first publicized by Yale University psychiatrist Harold Saxton Burr, who began studying the electromagnetic fields around animals and plants in the 1930s. He found, for instance, that the electromagnetic field of an adult salamander is present in baby salamanders, and may even be detected in salamander eggs.[38] Even though the egg is round, the outline of an adult salamander can be seen in the field. When a salamander limb is detached, the electromagnetic signature of the limb remains intact. This led him to other experiments in which he found fields around many other organisms. He also postulated, based on twenty years of experiments, that *diseases show up in this field a long time before they manifest as concrete pathologies.*[39]

Later studies with more recent research protocols, such as a 2004 study of c-reactive protein in angry people, support Saxton Burr's belief that we receive many signals before disease manifests. C-reactive protein is a stress-related protein; its levels rise dramatically in the body in response to inflammation, and it is considered a marker for future heart disease. The subjects in this particular study were all physically healthy adults who *did not show any symptoms* of heart disease. They were measured for levels of anger and depression using widely-accepted standardized tests. The researchers then compared the level of anger in the subjects, and the level of stress protein in their body. They found that the angrier subjects had elevated levels of stress protein in their systems, and were at higher risk for heart disease, even though no symptoms had yet shown up.[40]

If our system is flooded with stress hormones like adrenaline and cortisol for a few minutes, in response, for instance, to a near-collision with another car on the freeway, the incident quickly ends as a biochemical event. However, if we hold onto resentments and emotionally painful thoughts for extended periods of time, the very biochemicals that are meant to save us during an emergency become toxins.

Long-term exposure to cortisol and other stress hormones has a host of bad effects. It suppresses immune response, reduces bone formation, decreases muscle mass, and damages cells in the brain that can result in impaired memory and learning. If, on the other hand, we quickly release our stress and return to a biochemical baseline, we restore normal cellular operation. And that's vital to our longevity as well as our health. The same precursor hormone is used by our bodies to make both cortisol and DHEA. Just as cortisol has negative effects, DHEA has positive effects. It has "protective and regenerative effects on many of the body's systems, and is believed to counter many of the effects of aging."[41]

Harold Saxton Burr not only suggested that diseases show up in the patient's energy system before manifesting as symptoms; he also believed that physical diseases could be treated by restoring balance to the energy system.[42] A more recent experiment measured the electrical charge of the uterus of a group of women. The researchers found that the uteruses of women with uterine cancer had a negative charge, while those without cancer had a positive charge. The negative charge in many tumors is assisting in the diagnosis and treatment of breast cancer. It is also helping doctors to identify other forms of cancer, as well as other diseases.[43] Conduction is also different in healthy organs and unhealthy ones. When a current is passed through a healthy organ in one direction, the electrical resistance exhibited by the organ is measured. The electrodes are then reversed, and in a healthy organ, the resistance is the same. In an unhealthy organ, however, the resistance changes.[44]

Besides their value in diagnosis, electrical fields have great value in treatment as well: a meta-analysis of fifteen studies of the effect of electrical stimulation on the healing of chronic wounds found that, taken as a group, the wounds of the subjects exposed to electrical stimulation healed 144% faster than those that did not.[45] Indeed, one audacious series of experiments by French doctor Jacques Beneviste demonstrated the effects of electromagnetic fields in a novel way. Increased secretion of histamine increases heart rate. Rather than administering histamine itself, Beneviste simply exposed a beating heart to the *electrical frequency* of the histamine molecule, and by doing so speeded up the rate of contraction.

Next, the electrical signature of atropine (which *decreases* heart rate) was applied to the same heart. It reduced the flow of blood in the coronary arteries, just as the organic compound would have done.[46] The electromagnetic signatures of the compounds were having the same effect as the compounds themselves. At one point, to silence his many critics, Beneviste recorded the signals on a computer floppy disk in Paris, and mailed the disk to colleagues at Northwestern University in Chicago. After unwrapping the package and performing the experiments, the American researchers found that the effects of the signatures on the disk were "identical to effects produced on the heart by the actual substances themselves."[47] Other experiments have measured the electromagnetic signal of DNA molecules in a vacuum. The DNA matter is then removed. However, the vacuum still retains the imprint of the DNA's vibrational signature.[48]

Acupuncture is also an electrical treatment. A little-known fact about acupuncture needling is that it creates an electrical charge. It does so in three ways: The first is that acupuncture needles are stainless steel, with a handle made of a second, different metal. Two dissimilar metals, in the presence of an electrolyte solution like salt-water, generate an electrical current When placed in a salt solution, such as in human tissue, the two metals generate a small electrical current electrolytically.

After a few seconds in the body, the tip of the needle is warmer than the handle by as much as 25 degrees Fahrenheit. The differing thermal gradients stimulate a transfer of electrons from one to the other. The tip of the needle becomes positive relative to the handle. However, if the handle is heated or the needle is twirled— as acupuncturists often do—the tip becomes negative relative to the handle.

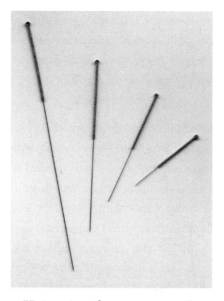

Various sizes of acupuncture needles

The insertion of the needle as it punctures the skin also creates a piezoelectric charge at the point of contact. When the acupuncture points on the energy meridians are stimulated, they may send piezoelectric currents through the meridians. Acupuncture points may in fact be portals of enhanced sensitivity through which entire meridians can be stimulated. In hundreds of clinical studies, acupuncture has been proven effective for a wide variety of ailments, from reducing the chest pain of heart patients who have been unresponsive to drugs,[49] to the restoration of fertility in men,[50] to the control of chronic tension headaches in around three quarters of patients.[51] Many Western practitioners now also hook up their acupuncture needles to devices that produce a controlled electrical current. A current of one cycle per second (1 Hz) passed through acupuncture needles has been shown to raise the level of a patient's beta-endorphins, molecules produced in the hypothalamus that aid in pain relief. [52]

Interestingly, the same effect can be obtained without acupuncture needles.[53] The massage technique of shiatsu stimulates acupuncture points with the practitioner's fingers, using pressure or tapping. Several Energy Psychology techniques have the doctor or patient tap on certain points with their fingertips, and with correct application are able to produce dramatic and immediate improvements in the patient's self-reported emotional state. Acupuncture points conduct electricity even when needles aren't used.[54]

Energy Medicine's Ancient Roots

While electromagnetic healing devices may be the newest application of energy in healing, the principles on which it is founded go back thousands of years. The ancient system of acupuncture arose out of the study of *chi* or *qi*, (literally "breath," a metaphor for "life energy" or what a modern scientist might describe as electromagnetic energy fields). Ancient Chinese medical practices, collected before the first century B.C.E. in a text attributed to the Yellow Emperor, described the cultivation of this life energy, or *qi*. It was thought to be enhanced by various physical postures, movements, and the conscious use of breath. The goal of these exercises was to achieve "a trance-like state of relaxation, wherein the *qi* can be regulated and directed by the mind to correct imbalances in the body."[55]

The principles of acupuncture may predate the Yellow Emperor by thousands of years. The summer of 1991 was one of the warmest in recent European history. During a hike in the Alps, two Germans tourists, Helmut and Erika Simon, came across what they first thought was the body of a hiker who had succumbed to the glacial ice, as had been the fate of several hikers in the previous few years. The Austrian authorities pulled the body loose from the ice, and only when it had been taken to Innsbruck for examination was it realized that this was no ordinary corpse. It turned out to be the mummified remains of an ancient man marvelously preserved in the ice. Radiocarbon dating showed the body to be from the period around 3300 B.C.E.

The study of the body, which researchers christened Otzi, has yielded many fascinating clues about the customs, dress, diet, and beliefs of the period. With him were found his copper axe, his bow and arrows, and dagger. The remains of his hide coat, grass cloak, leggings, shoes, and loincloth were with the body. Human blood from four different people was discovered on his weapons. A deep cut on his right hand led archeologists to conclude that he might have been in a mortal fight for survival, fleeing high into the Alps before expiring. Later examination revealed that an arrowhead had penetrated a major artery, and would have led to death in minutes.

Otzi was forty-six years old at the time of his death, and among his physical afflictions were intestinal parasites and arthritis. He also had some fifty-seven tattoos on his skin. Some were dark blue dots. Others were + signs, or short parallel lines. Infrared photography also revealed long tattooed lines. But the location of some of the tattoos is fascinating. They correspond with the meridian lines used in acupuncture to treat stomach complaints and arthritis, the very complaints from which Otzi is known to have suffered. The tattoos were either exactly on, or within a quarter-inch of, traditional acupuncture points and meridians.

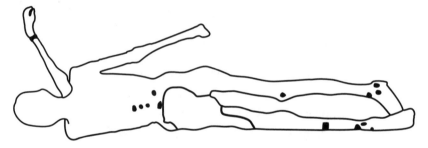

Position of some of Otzi's tattoos

It is thus evident that the locations of the meridians predates the Yellow Emperor's classic texts by thousands of years, and was known in prehistoric Europe, China, and probably India. Ancient sages and shamans developed knowledge of the flow of *qi* without any of the sophisticated technological measuring tools we use today. The Indian sage Susruta, writing around 1,000 B.C.E., described prosthetic surgery to replace limbs, caesarian sections, and rhinoplasty, cosmetic surgery of the nose. He was apparently an expert meditator. He is said to have been able to become so still during his meditations, so tuned to his body's signals, that he was able to chart the course of the blood flowing through his arteries, and draw an anatomically accurate rendition of their locations! He wrote a book called *Susruta Samhita,* one of the core texts of Ayurvedic medicine.

Human awareness of these energy channels along which *qi* flows is ancient indeed, and methods to enhance the flow may have been part of the knowledge base of prehistoric shamans from many cultures. Of the exercises designed to enhance *qi,* T'ai Chi is one of the most widely practiced in China and the West. It emphasizes large, gentle, slow motion movements, as opposed to the abrupt offensive movements of other martial arts. The Yang family's style of T'ai Chi, which is the most widely practiced, gained popularity in 1850 when the style's founder, Yang Luchan, was retained by the Chinese Emperor to instruct the Imperial Guard in the art.

Yang Luchan's grandson
Yang Cheng-fu demonstrating the Single Whip position

Adherents practice T'ai Chi by the thousands in parks and public spaces all over China in the mornings. Many believe that it boosts health and increases longevity, a view bolstered by a 2004 review of scientific studies of the subject in the *Archives of Internal Medicine*.[56] Clinical psychologist Michael Mayer, a modern Western Qigong master, in his book *Secrets to Living Younger Longer,* says that "States of consciousness are expressed in postures, and just as an actor practices 'stances' to enhance the expression of feeling, so does a Qigong practitioner practice his or her stance to maximize power, healing, and the expression of intention.... The intricacies of cultivating stances of power, healing and spiritual unfoldment have been matters of inquiry for Qigong practitioners over the last many thousands of years."[57]

Exercise is not associated with sacred awareness for most of us—how many times have you seen someone lifting weights while praying, or doing pushups with reverent mindfulness? Yet in many traditional cultures, exercise and the divine are linked. Constance Grauds, in *The Energy Prescription,* says that, "shamans (and yogis and Qigong Taoists) know spirit energy as the power of life itself, and view 'exercise' as a form of active communion with that power.... Sustainable exercise—exercise done with somatic awareness—may be the most powerful discipline for conducting spirit energy. It literally saturates our body with regenerating life force, from our muscles and bones to our very cells. And it takes our fitness and health to a whole new level."[58]

Mapping the flows of *qi* in the body for therapeutic purposes was a primary concern of these ancient Oriental doctors. They developed an elaborate diagnostic and prescriptive system, without the benefit of any of the tools we take for granted in the West today. It is a remarkable testament to their powers of observation that so many discoveries on the leading edge of modern scientific medicine are utilizing the same energy pathways they considered so important thousands of years ago. Whether activated by an exercise regimen like T'ai Chi, an electromechanical stimulation method like acupuncture, a biofeedback or EDS machine, or your belief system, the point of therapy is to restore full function and balance to the body's electromagnetic energy system.

Frequencies of Healing

The uses for electricity in medicine continue to expand. There are some 100,000 EDS machines in use worldwide, utilizing the electrical potential of acupuncture points for diagnosis and treatment.[59] According to the U.S. Department of Veterans Affairs, some 300,000 patients have permanently installed pacemakers.[60] By way of contrast, in 1959, the Elema 135 was the first pacemaker to be offered for sale. Total number of units sold that year? Two.[61]

There are tens of thousands of EEG, EKG, and other electromagnetic devices in use; the industry is now estimated to generate about $17 billion a year for the sale of its devices.[62] Transcutaneal Electrical Neural Stimulators (TENS machines)—some oscillating at the same frequency as human DNA—employ minute electric currents to relieve pain and increase electrical energy flow in the meridians. Devices using nano-currents that act at the level of a single cell are in development. We are calling in spirits that the ancient Ute, dancing around their ceremonial fires, illuminating the Colorado prairie with flashes from their quartz rattles, could never have dreamed of.

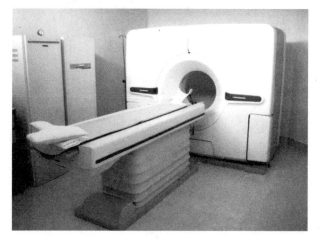

CT Scanner

In our book *Soul Medicine,* pioneering neurosurgeon Norm Shealy and I devote several chapters to the history and clinical applications of electricity and magnetism in healing.[63] In summary, the picture that is emerging of the mechanisms of healing utilized by these techniques is this: Electricity is generated by either the manual piezoelectric stimulation of certain points (acupuncture, acupressure, energy tapping), by distant electromagnetic fields, by quantum fields (nonlocal healing), or by an electrical device such as a pacemaker or a TENS unit. Any of these can produce a beneficial change in the body's electrical field and promote wellness. Dr. Shealy, who developed a state-of-the-art new TENS unit called the SheliTENS, describes its use in detail in his recent book *Life Beyond 100.*[64] He has collected many complete clinical medical histories documenting the healing of a variety of conditions using electrical stimulation, including hard-to-treat autoimmune diseases.

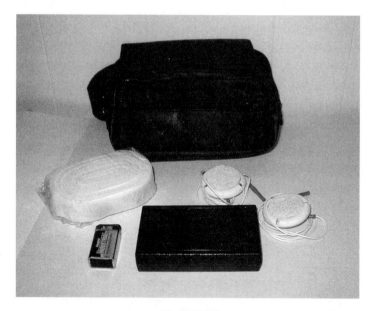

SheliTENS

Yet the fact that electromagnetic energy is all-pervasive in life does not answer many important questions. It does not tell us how this force travels, where it goes, or what it does. The answers to these questions lead to some exciting and very new discoveries in an entirely different field of human physiology, discoveries that have the potential to shake up medicine as dramatically as the overthrow of the dogma of genetic determinism.

5

The Connective
Semiconducting Crystal

Man is equally incapable of seeing the nothingness from which he emerges and the infinity in which he is engulfed.

—Blaise Pascal, 1623–1662

Connective tissue has not been perceived by the research establishment as a glamorous research target. You don't read breathless reports in the *Washington Post* of research published in the *American Connective Tissue Review*. While you might read a story about research published in the *American Heart Journal,* or *Neurobiology,* connective tissue has failed to generate the same level of excitement as other organs. This might be about to change dramatically!

You can think of connective tissue as being a system of bags that contains—and gives structure to—each of your organs, coupled with a system of wires that holds all the bags together. The wires wind through the joints of the skeleton, which supports the structure of wires and bags. Your organs are encased in sheaths of connective tissue called the fascia. The fascia surrounding muscles, called the *myofascia,* terminates in ligaments and tendons attached to the bone.

Connective Tissue System and Collagen

The tendons are composed of twisted collections of collagen bundles, each composed of collagen fibers. Collagen fibers are composed of collagen fibrils, assemblies of molecules secreted outside of specialized connective tissue cells, called fibroblasts. Taken as a whole, the connective tissue system is the largest organ of the body.

Below the skin and superficial fascia are the ligaments (white areas)

Yet the simplicity and ubiquity of the connective tissue system masks an important characteristic: connective tissue fibers are arranged in highly regular arrays. There is a name for a *highly regular parallel array of molecules,* whether it's in liquid or solid form: it's called a *crystal.* The collagenous molecules in which all your organs are encased function as a system of liquid crystals. Crystals—highly ordered arrays of molecules—are found in several different kinds of tissue, including:

The DNA in genes
The photosensitive rod and cone cells at the back of the eye
The myelin sheath of nerve cells

The collagen molecules that make up connective tissue
Muscle tissue's densely packed molecules of actin and myosin
The phospholipids of cell membranes

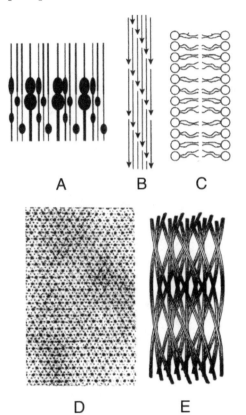

Tissues with highly regular arrays of molecules include (A) rod and cone cells in the eye; (B) collagen molecules in connective tissue; (C) phospholipids in a cell membrane; (D) cross-section of muscle tissue showing actin and myosin molecules, and (E) DNA molecules in a chromosome[1]

This crystalline structure of the collagen molecules that make up your connective tissue has a remarkable property: it is a *semiconductor*. Semiconductors are not only able to conduct *energy*, in the way the wiring system in your house conducts electricity very quickly from one point to another. They are also able to conduct *information*; think of your high-speed internet connection. Besides many other properties, semiconductors are also able to *store* energy, *amplify* signals, *filter* information, and to move information in one direction but not in another.[2] In other words, the connective tissue system can also *process* information, like the semiconductor chips in your computer. Your connective tissue system is well suited for the task of conveying both energy and information, because it connects every part of your body to every other part.

A connective tissue sheath (light color) surrounding blood vessels, nerve bundles, and muscle tissue

Think of the communication possibilities of this structure: *every organ of your body is encased within the body's largest organ, which functions as a liquid crystal semiconductor* in the form of the connective tissue system. Another property of connective tissue is that it is a piezoelectric substance, which when compressed, generates electricity. "The piezoelectric constant of a dry tendon, for example, is nearly the same as that for a quartz crystal"[3]—the same kind of crystal the Ute Indians used to create light in their sacred rattles.

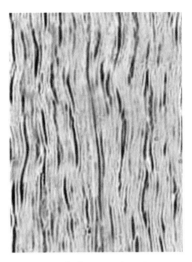

Magnified connective tissue

Semiconduction in Connective Tissues

You can see why it's time for us to start getting excited about the connective tissue system! The importance of its conductive property is hard to overstate, for it explains one of the enduring mysteries of acupuncture: how stimulating one acupuncture meridian point, for instance in the ear, can have an effect on another point, for instance in the spleen. Mae Wan Ho, a researcher who has published several papers on the crystalline nature of connective tissue, says, "Liquid crystallinity gives organisms their characteristic flexibility, exquisite sensitivity, and responsiveness, thus optimizing the rapid, noiseless intercommunication that enables the organism to function as a coherent, coordinated whole."[4]

Intercommunication also explains how tapping one part of the body, which creates a piezoelectric signal, then creates an effect in a distant part of the body, or in the entire body. Electrical signals can be carried throughout the body by the connective tissue. James Oschman, Ph.D., whose fascinating book, *Energy Medicine in Therapeutics and Human Performance*, assembles a great diversity of research from many disciplines into an understandable whole, says, "Signals generated by the piezoelectric effect...are essential biological communications that 'inform' neighboring cells and tissues.... The fully 'integrated' body may be a body that is entirely free of restrictions to the flow of signals."[5]

Cell biologists studying collagen reductively would miss these unique properties. Broken down into individual collagen molecules, connective tissue does not have the same characteristics; it takes its collection into a *parallel array structure* to produce its ability to conduct and store energy. The electrical properties of the connective tissue also explain how cells can communicate much faster than the speed of neural transmission.

Speaking into Cells

Each cell also has a skeletal structure. The shape of the cell is partially determined by a system of cylindrical protein "beams" called *microtubules*. Microtubules are the girders of our cells. They are rigid protein structures—the rebar that give cells their shape.

Microtubule bundles in the limb of a simple organism

The unglamorous role of being merely a member of the "supporting" cast has resulted in less attention being paid to microtubules by researchers, who have had much more glamorous cell structures—such as the mitochondria (the energy generators), the genes (repository of blueprints), and the cell membrane (letting signals in and out)—to occupy their attention. Being a piece of rebar in a building is unlikely to attract as much attention as being the facade, the power plant, the lobby, or the elevators.

We think rebars and we automatically assume rigidity. Microtubules do indeed provide cells with the property of rigidity. One of the overlooked properties of microtubules, however, is their transience.

Microtubules are transient in the sense that they are rebuilt often—in some cells, several times an hour. Far from being structures that are created during the splitting of a cell—and remain firmly in place throughout that cell's life—microtubules are rebuilt often, and quickly. In the June 2004 issue of *Science and Consciousness Review,* John McCrone writes this about the microtubules of the brain: "Do you know the half-life of a microtubule, the protein filaments that form the internal scaffolding a cell? Just ten minutes. That's an average of ten minutes between assembly and destruction.

"Now the brain is supposed to be some sort of computer. It is an intricate network of some 1,000 trillion synaptic connections, each of these synapses having been lovingly crafted by experience to have a particular shape, a particular neurochemistry. It is of course the information represented at these junctions that makes us who we are. But how the heck do these synapses retain a stable identity when the chemistry of cells is almost on the boil, with large molecules falling apart nearly as soon as they are made?

"The issue of molecular turnover is starting to hit home in neuroscience, especially now that the latest research techniques such as fluorescent tagging are revealing a far more frantic pace of activity than ever suspected. For instance, the actin filaments in dendrites can need replacing within 40 seconds...[and]...the entire post-synaptic density (PSD)—the protein packed zone that powers synaptic activity —is replaced, molecule for molecule, almost by the hour."[6]

The conclusion of this research is that the entire brain is being recycled about once every other month, opening up an enormous field of enquiry into how neurological change interacts with changes in energy systems.

The very skeleton of each cell is being revised several times each hour, as though a building's architecture were metamorphosing each day! Far from the role of the passive supporters to which microtubules have sometimes been assigned, it turns out that they are active daily shapers of the very physical dimensions of a cell's structure. Cutting-edge scientists are left wondering not so much how we can change, but how we can endure.

Yet this is only the beginning of our reevaluation of a microtubule's role. For it turns out that their reformulation of the structure of the cell is not random. Microtubules appear to be *attuned to the energy field in which the cell exists.* Not just mere girders, microtubules may in fact also be antennas, receiving signals from the environment and carrying out their complex restructuring in response to these signals.[7] The cytoskeletons of cells are also piezoelectric conductors; "electric fields change their shape, and mechanical stimuli their electrical condition."[8] The role of the hollow core at the center of a microtubule is unknown, though some researchers have speculated that "superconductivity and electromagnetic focusing take place in the cores."[9] Cell biologist Bruce Lipton points out that "Conventional medicine works with the iron filings, whereas a deeper form of healing would attempt to influence the magnetic field. Most doctors don't see the field, so they're trying to figure out the relationship between the filings without even trying to incorporate the energy field in which they exist."[10] Microtubules may be a key resonator with that energy field, and some researchers have suggested that they may play a key role in receiving information required to structure cells from the "conductor" of consciousness.[11]

Anchoring Cells to the Connective System

An important class of protein sugars, called the *integrins*, conveys information between the environment outside the cell and the inside of the cell. They anchor the cytoskeleton and other internal elements of the cell's structure—even the nucleus—to the connective tissue outside of the cell. So there is a feedback loop, of information going into the cell and information emanating from the cell and being conveyed into the connective tissue matrix.

The implications of this correspondence are staggering. It turns out that the internal structures of cells are also influenced, through the connective tissue, with the energy environment we are creating with our thoughts and feelings. It is not just DNA that is being affected by our internally produced environment; it is not just our collagenous tissues; it is the very microtubular structure of our individual cells. *Our bodies are more plastic than we once thought, and this flexibility carries with it the possibility of rapid, miraculous healing.*

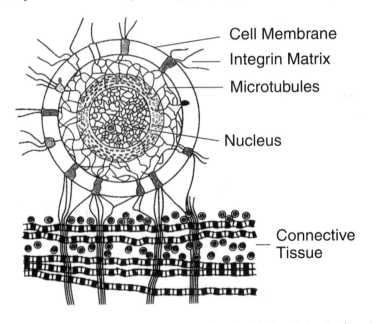

Tissue microstructures showing connections from inside the cell (nucleus) to the environment (connective tissue)[12]

The chemical signaling mechanisms of the nervous system communicate at a speed of between 10 and 100 meters per second, depending on the diameter of the axon. This is much too slow to explain many human processes. The neural signaling mechanisms described in most medical textbooks today are inadequate to explain, for instance, how a batter can hit a baseball. At a speed of 50 meters per second, there is simply not enough time for the visual image of the ball leaving the pitcher's arm to travel from the batter's eye, through the optical nerve, to the visual cortex, then to the centers of the brain that govern volition, then to those that generate muscle movement in the upper body, then down to the muscles themselves, instructing them to swing. There is only about two-fifths of one second between the ball leaving the pitcher's hand and the completion of the batter's swing.[13] Neural signaling cannot do the job of conveying such a rapid response. Baseball catcher Tim McCarver rightly exclaimed, "The mind's a great thing as long as you don't have to use it."[14] This is just one of many functions in higher animals that cannot be explained in terms of the capabilities of the neural signaling system.

Each cell is going through around 100,000 chemical processes per second. These processes are coordinated among a body—a community of cells—that is composed of trillions of cells. Such coordination borders on the miraculous. How do these cells know how and when to create exactly the same reactions at exactly the same instants? In *The Field,* Lynne McTaggart asks: "If all these genes are working together like some unimaginably big orchestra, who or what is the conductor? And if all these processes are due to simple chemical collision between molecules, how can it work anywhere near rapidly enough to account for the coherent behaviors that live beings exhibit every minute of their lives?"[15] Neural signaling cannot even begin to grapple with a cybernetic task of this magnitude.

Rapid Cellular Signaling

If, on the contrary, we assume electromagnetic signals passing through a liquid crystal semiconductor encasing all the organs, with the microtubules in individual cells resonating in sympathy, many such processes suddenly make sense. Oschman sees the connective tissue system as a living network that corresponds to the operating system of a computer, enabling all the parts to work together smoothly.

Joie Jones, Ph.D., and his research associates have tested the results of stimulation of the vision-related meridian points in the foot, using an fMRI imager. They discovered that when these points are stimulated, neural circuits in the brain's occipital lobes are activated almost immediately—at a rate far faster than neural conduction can explain.[16] In the real world, our bodies function with an integration of stimulus and response at a speed much faster than that of neural signaling. A group of researchers led by Andrew Ahn, M.D., of Harvard Medical School looked at the possibility that "segments of acupuncture meridians that are associated with loose connective tissue planes (between muscles or between muscle and bone)" had greater electrical conductivity than non-meridian points, and demonstrated this to be the case in a series of experiments published in 2005.[17] Signals travel through these electromagnetic conduction pathways in the meridians at a pace many orders of magnitude higher than the signals traveling through the neural net.[18]

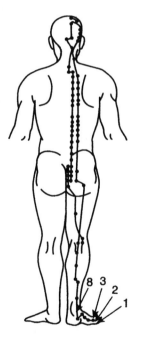

Stimulation of the acupuncture meridians in the foot associated with vision (8, 3, 2, 1) affects the visual cortex in the brain far faster than speed of neural transmission can explain[19]

Just like my brain sending a signal to my body, I can send a signal to another trusty assemblage of functionality—my car—that I want to unlock it and drive home. I can do this in one of two ways. One of them takes a long time, the other occurs instantaneously. Attached to my key ring is a car key. I can insert it in the car door lock and turn it, then open the door, reach inside to the other doors, and pull up on the door knobs to unlock them, too. Then I can insert the key in the ignition, turn it, and start the car. The process takes me about six seconds, and is accomplished by a physical-mechanical signaling system.

I can also use the other element attached to my key ring, the automatic door opener and engine starter. One click, and, presto, the doors are unlocked and the engine starts. It takes a nano-second to start the process, which is accomplished via an electromagnetic signaling system.

Another example of a physical signaling mechanism is a letter. I can seal it in an envelope, drop it in a mailbox, and a mechanical system of trucks, hands, and conveyor belts will deliver it to the addressee in a few days.

Or I can send the same message using an electronically based system. I can write the same letter, then simply hit the "send" key on my e-mail program. The communication reaches the recipient almost instantaneously. Psychologist David Feinstein, Ph.D., says, "Electromagnetic frequencies are a hundred times more efficient than chemical signals such as hormones and neurotransmitters in relaying information within biological systems, a calculation based on research conducted in the 1970s by Oxford University biophysicist C.W.F. McClare."[20] This is not terribly surprising when you consider that many of the body's regulatory chemicals travel less than a centimeter in a second while an electromagnetic wave could have traveled three-quarters of the distance to the moon in that time!"[21]

Your body has both systems: a mechanical-chemical signaling system, based on the movement of charged ions across cell membranes, and the diffusion of hormones and neurotransmitters; and an electromagnetic signaling system. Both the mechanical and the electromagnetic system can activate cells and genes to accomplish the intent of the user. Yet current medical practice focuses almost exclusively on the former, and gives an inexplicably small amount of attention to the latter. And conventional medical treatment protocols practically ignore the much faster and more efficient energy signaling system.

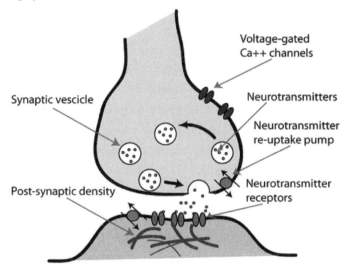

Voltage-gated
Ca++ channels

Synaptic vescicle

Neurotransmitters

Neurotransmitter
re-uptake pump

Post-synaptic density

Neurotransmitter
receptors

Conventional biomechanical/chemical signaling option

Biomechanical option (key) vs. electromagnetic option (button)[22]

A field of electromagnetic radiation as the cause of the structure of the body and its cells was first postulated by Russian scientist Alexander Gurwitsch as early as the 1920s,[23] in the same decade that the energy field of the heart was being mapped by Willem Einthoven and the EEG was being developed by Hans Berger. Also in the 1920s, researcher Elmer Lund at the University of Texas found that he could reorganize the cellular structure of small animals by applying an electric current strong enough to override the creature's electro-magnetic field.[24] Later studies by Herbert Fröhlich of the University of Liverpool predicted that liquid crystalline phospholipids just below the cell membrane, vibrating at certain frequencies (now called Fröhlich frequencies) could "synchronize activities between proteins and the system as a whole."[25] He showed that at certain energy thresholds, "molecules begin to vibrate in unison, until they reach a high level of coherence. The moment molecules reach this state of coherence, they take on *certain qualities of quantum mechanics, including nonlocality.* They get to the point where they can operate in tandem."[26]

The semiconductive properties of the connective tissue acting in resonance have a speed and power that far transcends other sig-naling mechanisms. Mae Wan Ho says that the crystalline structure of tissues and organs results in harmonic resonance of the entire structure: "When the coherence builds to a certain level...the organ-ism behaves as a single crystal. ...A threshold is reached where all the atoms oscillate together in phase and send out a giant light track that is a million times stronger than that emitted by individual atoms."[27] This quantum coherence means that signals are passed back and forth, in a continuous feedback loop, from cells to brain to tissues, in virtually the same instant.

Communication is a two-way street. Not only does the environ-ment of our thoughts and feelings communicate itself throughout the body via our liquid crystal semiconductor system, but changes in state of our semiconductor system communicate themselves instantly to our consciousness. Our brains are receiving information from all over our bodies rapidly and continuously, and using this information to shape our choices, both conscious and unconscious. Physiology and consciousness are in an ongoing and inseparable feedback loop, rather than being two separate processes. Oschman calls this the *liv-*

ing matrix, and summarizes by saying that, "The connective tissue and cytoskeletons together form a structural, functional, and energetic continuum extending into every nook and cranny of the body, even into the cell nucleus and genetic material. All forms of energy are rapidly generated, conducted, interpreted, and converted from one to another in sophisticated ways within the living matrix. No part of the organism is separate from this matrix."[28] The resonant nature of the living matrix also suggests that it is possible that *our connective tissue system may be a quantum resonator,* conducting signals from the quantum field of the universe into the body, and from the body to the field.

The signals from our brains are communicated via the matrix of energy conduction within our bodies constantly. Every thought you think is echoing through your connective tissue communication system, turning genes on and off, producing either stress responses or healing responses. This understanding opens up a vast new panorama of potential self-healing.

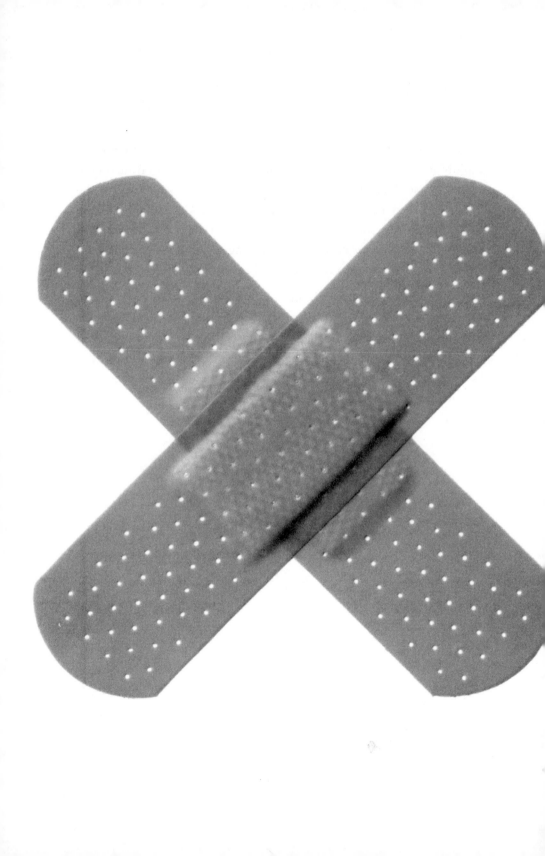

6

Anatomy of a Monster

Here's a wrenching fact: If the United States had an infant mortality rates as good as Cuba's, we would save an additional 2,212 American babies a year.

—Nicholas Kristof, Pulitzer prize-winning journalist
New York Times, Jan 14, 2005

Medicine in Western society is in crisis. Many informed people will tell you so: doctors, hospitals, insurers, politicians, alternative medicine practitioners, and the makers of public policy. But they are almost all wrong. They are wrong about the nature of the crisis, wrong about the causes of the crisis, and wrong about the solutions to the crisis. The debate about "health care" is everywhere: in magazines, in boardrooms, in debates between political candidates, on web forums and the electronic media. It fills thousands of minutes, lines, and bytes—yet all this chatter manages to miss the most important problems—and their solutions.

Take the figure—much-quoted in U.S. media—that 46 million Americans (some 12% of the population) do not have health insurance.[1] Getting these uninsured people into "the system" has dominated public debate for two decades. It was to be the social centerpiece of President Clinton's first term, until it hit the opposition of entrenched interests and floundered. It is a prime talking point in every election.[2] Yet it's entirely the wrong direction for public policy to go: the system into which politicians want to bring the uninsured manages the remarkable simultaneous feats of incompetence at keeping people well, and spending obscene amounts of money to accomplish this.

The costs are truly staggering. The U.S. spends about three times as much as Great Britain per person on health care, and twice as much as Canada.[3] Both those countries have universal health care—every single person is covered. The U.S. spends almost twice as much as Germany and France, and more than twice as much as Japan.[4] These are all comparable post-industrial societies and economies; some, like France, have inefficient, fossilized, or costly systems. If American costs were 20% higher than the average among rich industrialized countries, its citizens would have cause for concern. Costs that are more than double the average are a national disgrace. And that's for a system that does not cover everyone.

Solutions like capping payments to doctors in the Medicare program that provides care to the elderly, or moving patients into Health Maintenance Organizations, have, at best, slowed the growth of spending. They have not brought costs down. Medicare is currently predicted to go broke by the year 2019.[5] The unfunded liabilities of the system for the next seventy-five years have been calculated at a staggering $27.7 trillion.[6] Faced with this crisis, national leaders have recently not been content with merely ducking the issue. They've made it worse by piling on added benefits for seniors (who vote in large numbers), that today's children (who can't vote at all) will have to pay for—with interest. A recent special report by a panel of economists concludes that current public policy is so misguided that it "may speed the reform of American health care, but only by hastening the day the current system falls apart."[7]

If this bizarre amount of spending were producing a marvelously healthy population, there might be some defense for it. It isn't:

- Infant mortality rates are higher, and life expectancy lower, than every single one of the countries mentioned above.[8]

- According to a large-scale study of twelve different metropolitan areas from Newark, New Jersey, to Miami, Florida, Americans get substandard medical care more than half the time, leading to "thousands of needless deaths each year."[9]

- We're both stressed and depressed; according to the one study, "About one in five Americans now suffers from a diagnosable mental disorder."[10]

- A study entitled "Trial Lawyers, Inc.," estimates that, "the total costs, direct and indirect, of health-care litigation (including suits against doctors, drug firms, HMOs, nursing homes and so on) could be as much as $200 billion—a Hurricane Katrina every year."[11]

No wonder Andrew Weil, M.D., begins an authoritative survey of the medical profession, published in *The Archives of Internal Medicine,* with these words: "The chassis is broken, and the wheels are coming off."[12]

Asking questions like, "How can we slow the growing cost of our medical system," "How can we pay for the Medicare prescription drug benefit," and "How can we bring the uninsured into the system," have taken public debate in entirely the wrong direction. They are the wrong questions. As long as pundits keep on asking them, they will get the wrong answers. If Western society sticks with its disease-centered and money-driven model, citizens' health will not improve, even as they continue to pay more. Good ideas are what we need; when our thinking changes, our institutions change. Changing the institutions without changing the flawed thinking behind them results only in hastening their collapse, as was evident in the Medicare prescription drug benefit passed by Congress in 2004; it brought forward the date of Medicare's insolvency by eight years.[13] Asking how to get the uninsured into the system—which is as far as most policymakers think—is like asking how to cram more passengers onto a train that is already jammed, has some cars that are falling to pieces while others are trimmed with diamonds, is about to crash, and is heading in the wrong direction anyway.

What, then, are the right questions?

Let's start with, "How can we make the largest number of people as well as possible?" This question abandons our old way of thinking about how to tinker incrementally with our broken system, and invites a fresh awareness. Our assumption shifts away from treating disease, and toward creating health. As we create health, there is less disease to treat. This approach assumes that consciousness is an essential part of the equation, and that beyond the mere absence of symptoms, high-level wellness is the goal.

What can we teach every high school student, every worker, and every retiree that would maintain their bodies and minds in the best possible condition for the longest possible time? According to the Centers for Disease Control, lifestyle choices have a greater health effect than medical care.[14] And once their bodies are functioning well enough to circumvent the distraction of illness or low vitality, how do we optimize their spiritual and emotional wellbeing? There are four simple and obvious remedies. They cost almost nothing. They require very little time and attention; much less attention, certainly, than being sick does. And if implemented, they would radically alter the health picture of our entire civilization within thirty days.

Here's a baseline that every person could be supported in achieving:

- A period of aerobic exercise, flexibility, and strength training averaging just twenty minutes a day.

- A diet based on protein and complex carbohydrates.

- Supplements: at minimum, a capsule each day of multivitamins, antioxidants, and fish oils.

- Proficiency in a *chronic* stress-reduction technique such as meditation, prayer, or contemplation, plus an *acute* Energy Psychology intervention for use in crises.

The first three items provide a foundation of physical and mental health on which a structure of spiritual wellbeing can be built. It's hard to meditate, do yoga, contemplate, or pray with serenity when your body is sick, so addressing basic physical wellbeing is an early-stage requirement. The health benefits of meditation alone are well documented and numerous.

Meditation has been shown to lower blood pressure, improve resting heart rate, reduce the incidence of strokes, heart disease, and cancer, diminish chronic pain, ameliorate anxiety and depression, and have a beneficial effect on many other diseases.[15] If meditation were a drug, it would be considered medical malpractice for a physician to fail to prescribe it. The results of studies of prayer are equally impressive.[16] A single brief period of spiritual and emotional centering in the morning positively affects our immune system all day long[17] and sets us up with a healthier and more peaceful emotional baseline. To cope with sudden stresses that disrupt our baseline, we need an additional technique drawn from Energy Psychology to quickly discharge the disturbed emotional energy and allow us to return to a peaceful, relaxed state.

The answers to our dilemmas are staring us in the face. Millions of ordinary people in Western societies are bucking the national trend towards obesity and illness by using one or all of these simple remedies. They are daily countering the dysfunctionality of the disease- and money-centered medical paradigm. If a majority followed the simple four-part prescription above, which *takes less than an hour* per day and *costs less than one dollar* per day, much of the superstructure of society's current system would become obsolete. The numbers of people requiring treatment would drop precipitously. As far fewer people utilized the system, costs would fall, creating a pool of funds to treat all those, including the uninsured, who require the ministrations of medicine, whether it's conventional medicine, energy medicine, or complementary medicine.

Future Medicine: Fast, Cheap, Effective

Not only are such interventions cheap and effective—they are fast. Recent studies show that fish oil supplementation leads to marked improvements in mental acuity in thirty days or less. Fish oils have been shown to prevent cancer[18] and heart attacks, improve cardiovascular health,[19] and reduce the incidence of diabetes, autoimmune diseases, and inflammatory diseases.[20] Some of the ingredients in antioxidant and multivitamin tablets are absorbed by the body in minutes. Many of the benefits of exercise and meditation occur immediately, others show up cumulatively in the form of stronger, more durable bodies, and a reservoir of inner peace from which people who pray and meditate can nourish themselves in times of stress. The sense of physical and spiritual wellbeing that people get from these simple lifestyle choices can improve their lives dramatically in a month. When we start to ask the simple question about how we support ourselves in optimal wellness, we are led by compelling data to the four simple lifestyle choices above. Science now points us inexorably toward the value of diet, exercise, meditation, and wellness-based therapies.

Shifting the health habits of an entire civilization seems like a tall order. The system towers like Godzilla over Western society, devouring social resources, impervious to any change except one that strengthens it. This frozen system affects every individual. People wishing to attend a yoga class, join a gym, get a massage, visit an alternative healer, or otherwise nurture their wellness must pay for it themselves. Dr. Dozor observes, "To health insurance companies, integrative healing practices appear to be simply another expense."[21] Dean Ornish, M.D., ponders, "How did we get to a point in medicine where interventions such as radioactive stents, coronary angioplasty, and bypass surgery are considered conventional, whereas eating vegetables, walking, meditating, and participating in support groups are considered radical?"[22]

Yet I predict that our definition of what is—and is not—conventional in medicine will shift, and shift quickly. Scholar Jean Houston, in her book *Jump Time,* points out that social evolution does not happen in a gradual upward curve.[23] It is marked by long plateaus, followed by rapid jumps. She identifies the Renaissance as one such jump. Within 25 years, all the assumptions of society had changed—assumptions about politics and governance, about money and economics, about gender roles, about religion and science, and about health and medicine. This revolution was brought about by a tiny number of people—perhaps a thousand in all. She believes that we are in the middle of another such jump, and that the landscape of consciousness will be radically different by the second and third decades of the twenty-first century. Health-conscious Westerners today are jumping with their pocketbooks. The dollar amount spent on alternative therapies now exceeds Americans' out-of-pocket payments for conventional medical care, according to Dean Ornish,[24] though that may change as companies require that workers pay higher deductibles. Comparable figures for other Western countries are not readily available, but the proliferation of alternative treatments in Europe, and the huge increase in the number of practitioners, suggest that similar trends are occurring throughout Western society.

The Psychosphere of Health

The benefits of getting and staying healthy are so obvious that they are attracting social awareness like a magnet. As I have surveyed the field of wellness for several large-scale anthologies, interviewing dozens of authorities on the leading edge of wellness research, I have been struck by two things. One is that *our crisis is not one of money, access, or technology.* It is a crisis of *consciousness.* As we reformulate our social debate in order to ask the right questions, as we change our collective minds, better answers will emerge as day follows night. All our efforts to change "the system" won't have nearly the effect that a change of heart and mind will have on all our systems, both global social systems and personal physiological systems. Social consciousness, that which Teilhard de Chardin referred to as the "psychosphere," is the aggregate of the consciousness of all the individuals in the society. When many individual consciousnesses shift, the group consciousness inevitably shifts to reflect the new mix. Personal inner change is, in aggregate, outer social change.

As we as individuals begin to set aside time in our busy days for meditation, prayer, and contemplation, we become aware of the immanent side of existence. We become attuned to the flow of energy through our bodies, and we become more sensitive to the power of intention to shape our lives. As well as tackling our challenges by outer action, we awaken of the power of pure awareness. As inner peace and serenity become an integral part of our daily practice of life, we become aware of the infinite possibilities of the electromagnetic energy fields in which we have our being. We see how changes in energy can result in changes in outer form, and how changes in our awareness can affect our world. From a calm mind and a serene heart, and a sense of being part of a larger ordered whole, we see new potentials for our lives and make choices that unlock those potentials. We live transcendent lives.

The results are good for our bodies. The effects of spiritual or religious practice and social support on people undergoing heart surgery were studied by a group of researchers from the University of Texas Medical School, led by Thomas Oxman, M.D. They discovered that those patients that were richly networked or had a strong spiritual or religious practice had only one-seventh the mortality rate of those who did not.[25] These astonishing results are not an isolated instance. The effect of consciousness upon physical healing is already documented, and is being studied more and more. Larry Dossey points out that there are some 150 studies already completed, with more in progress.[26] This improvement in physical health gives an individual greater energy and vitality, which leads to more attention being available for meditation, yoga, prayer, and other spirit-strengthening activities. Integrating these activities into one's life leads to better physical health, in a virtuous and mutually reinforcing cycle.

Health care will look very different after the anticipated jump. Clinics will be places of spiritual renewal and emotional restructuring, rather than service stations for mechanical defects in bodies. Science will be welcomed as the ally of holistic approaches, rather than merely a tool of drug companies and medical technology manufacturers. Doctors and patients may routinely pray together, and pray for each other. Patients' mental, emotional, and spiritual states may receive as much of a workup as their bodies, if not more so, and intention and other intangible factors might become the primary method of treatment. People will be trained to use effective self-interventions as the first line of treatment. Many different healing modalities may be combined, with the line between what we now call "conventional" care and "alternative" care blurring into the question: "What combination of approaches will help us to be most well?"

I observe a collective social consciousness in the process of rapid, radical, and irreversible change, a change only thrown into sharp relief by counter-indications like the alarming increase in obesity. Public society and governance are still mired in the wrong questions, about bringing the uninsured into the system and about limiting cost increases. But enough individuals are finding personal answers that, in aggregate, may topple the seemingly immovable statue of conventional medical care within the next few years. Margaret Mead famously observed: "Never doubt that a small group of thoughtful, committed people can change the world. Indeed, it is the only thing that ever has." Every day that any of us prays, meditates, exercises, sets intentions, eats a healthy diet, and maintains a calm state of mind is a day in which the fossilized old edifice of conventional medical care is given a little nudge. One day soon, the aggregate of those nudges may bring it crashing down into the sea of common sense, out of which a new health care model is right now being born.

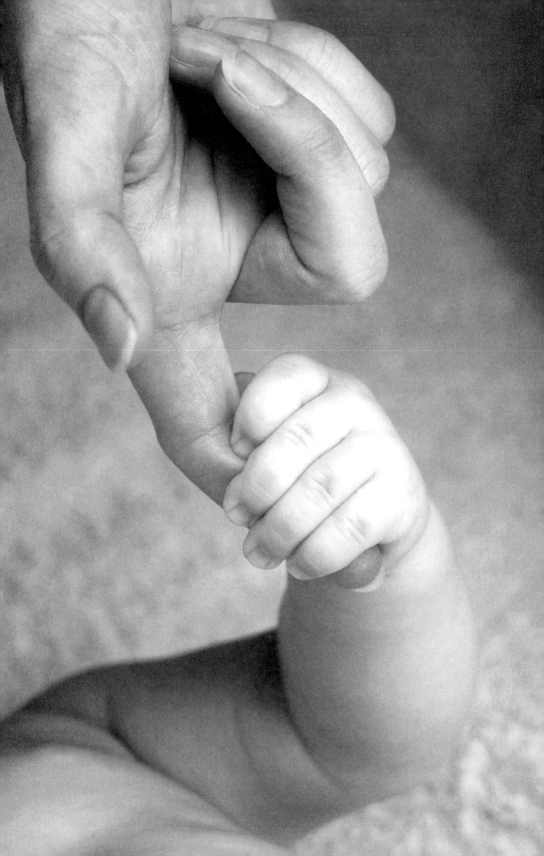

7

Consciousness as Medicine

A scientific worldview which does not profoundly come to terms with the problem of conscious minds can have no serious pretensions of completeness. Consciousness is part of our universe, so any physical theory which makes no proper place for it falls fundamentally short of providing a genuine description of the world.

—Sir Roger Penrose, *Shadows of the Mind*

O
ne of the most striking features of research on consciousness in healing is the size of the effect. If a new drug or surgical treatment for angina produced a statistically significant improvement in the health of patients of 15%, it would warrant a feature story in the *New York Times*.

Some of the studies of meditation, prayer, spiritual practice, and social goodwill and other intangibles sometimes show effects that are much more dramatic—though they are rarely publicized in the mainstream media. Some studies of spiritual practices and sociable behavior show that they produce a *sevenfold* reduction in mortality.[1] It is hard to imagine any conventional medical treatment that comes close. Intangible changes in consciousness, as well as having potentially huge effects, are also usually free, have no side effects, and do not require the patient to become enmeshed in an expensive, dysfunctional, and even dangerous medical "system."

Besides the size of the effects, the pervasiveness of the benefits of consciousness change is becoming apparent. Many studies, going back over thirty years, show changes in brain function among meditators. Newer studies are demonstrating that meditation does not just make subjects calmer; it has beneficial effects on many different measures of their health. A recent study of mindfulness meditation by researchers at the University of Wisconsin at Madison showed that it produces significant rises in a variety of antibodies and blood cells—all associated with increased immune function. This new crop of studies is showing that in addition to the other benefits of meditation, it measurably improves the body's ability to resist disease and the effects of stress—throughout the many organs and systems of the body.[2]

Mitchell Krucoff, M.D., of Duke University, and Suzanne Crater, R.N., his nurse research assistant, refer to the effects of consciousness as *noetic* effects. One of their experiments measured the amount of worry being experienced, prior to treatment, by patients who had been admitted to hospital because of unstable coronary syndromes. The period before treatment, especially for a known killer like cardiac disease, is likely to be filled with anxiety for a patient who has just been admitted to a coronary intensive care unit. This stress is not conducive to effective treatment, whatever that treatment turns out to be. The researchers wanted to find out what effect their noetic interventions might have on such patients, and whether it would reduce their worry. Stress management, touch therapy, and imagery were done in a single thirty-minute session just before treatment. The researchers measured factors like the patient's sense of happiness, satisfaction, fear, worry, calmness, and shortness of breath.

The patients in the experimental group all experienced a reduction in their sense of anxiety, regardless of the noetic intervention performed.[3] Interestingly, in a follow-up to this study done after the treatment, the patients who had reported a decrease in worry and an increase in hope after the treatment, also showed these same characteristics six months later.[4] This is a profound long-term beneficial effect to result from a mere single thirty-minute treatment!

Changing Our Medical Minds

The medical paradigm in the Unites States has gone through several stages. The first stage was roughly parallel to the century up until the First World War. The 1800s were more than a period of exploration of the Western frontier; they were also the Wild West of medicine in which dedicated practitioners mixed with snake-oil vendors, experienced Civil War combat surgeons rubbed shoulders with "doctors" whose only qualification was the cedar shingle hung from their front porch. During that period, the clear-sighted clinical observations of such geniuses as Andrew Still, the founder of Osteopathy, were indistinguishable to the layman from the extravagant claims made for patent elixirs and technologies promising, "Be a Radioesthesiast! Only $9.99—includes instructions!"

That First Stage, the Wild West of Medicine, was tamed by the Flexner Report, commissioned by the Carnegie Foundation and published in 1910. It ushered in the Second Stage by setting up licensing and professional standards for medical schools—or at least the AMA-sponsored institutions it favored. This new rigor was welcome in that it led to the outlawing of the worst kinds of quack medicine.

Unfortunately, it also led to the censure of many of the most original and effective "unconventional" treatments. It also fostered the rapid ossification of medicine into an orthodoxy controlled by the American Medical Association. For instance, by 1918, all but one of the homeopathic medical colleges—up to then the main competition to conventional medicine—had been forced to close their doors. Practices such as Osteopathy, and later Chiropractic, struggled to survive. The second stage might be considered the Age of Conformity in Medicine, when any promising findings outside of officially sanctioned channels—like those of Royal Rife in pathology, Harry Hoxsey in oncology,

Wilhelm Reich in psychiatry, and D. D. Palmer in skeletal manipulation—were suppressed or their founders punished. "Because physicists had not yet discovered the quantum universe," notes a modern cell biologist, "energy medicine was incomprehensible to science."[5]

The priesthood of biomedical orthodoxy came under increasing strain in the second half of the twentieth century. In 1987, chiropractic won a victory when federal judge Susan Getzendanner ruled that the American Medical Association had illegally conspired to destroy the profession through the restraint of trade. New discoveries and the anecdotal evidence reported by patients and physicians formed a background conversation that became increasingly difficult for the mainstream medical establishment to ignore.

Using the Mold to Break the Mold

Randomized controlled double-blind studies, the gold standard of medical research, have been upheld by the medical establishment for more than four decades to determine the efficacy of new drugs and surgical techniques. Ironically, that same scientific method, when it began to be applied to alternative therapies in the 1980s and 90s, began to turn up results like the studies presented here, ushering in the Third Stage, the Age of Infinite Potentials.

The characteristic of this stage is that the supposed antagonism between "conventional" and alternative medicine is being shown for the hoax it always was. Emblematic of this shift is the irony that the same scientific tool—clinical studies—that had allowed the drug companies and surgical instrument makers to capture the high ground of medical treatment during the Age of Conformity, now became the means by which many useful alternative therapies were shown to be far more effective than their conventional counterparts. The world of medical research is being turned on its head. The studies in this book herald the dawning of the Age of Infinite Potentials, in that the scientific method is validating the efficacy of treatments that would have been laughed out of a consulting room or professional convention twenty-five years before. A century after it began to transform physics, quantum theory has hit medicine.

The "infinite potentials" part of this age is that the potential of some of the alternative therapies is only starting to become apparent. We may need to adjust our thinking about the upper limits of human longevity, the duration of the human health span (as opposed to "life span," the term "health span" is the number of *healthy* years lived), the scope of brain function, the implications for medicine of quantum physics and especially the phenomenon of non-locality of consciousness, and the upper limits of human physical performance.

Recently, Jay Olshansky, Ph.D., professor of epidemiology and biostatistics at the University of Illinois at Chicago, debunked the idea that human life expectancy might increase significantly in the years to come. In a *New York Times* article that holds that, "the era of large increases in life expectancy may be nearing an end," he asserts that, "there are no lifestyle changes, surgical procedures, vitamins, antioxidants, hormones, or techniques of genetic engineering available today with the capacity to repeat the gains in life expectancy that were achieved in the twentieth century."[6]

I believe that this conclusion could not be more wrong. It is based on the extrapolation of Second Stage science. Third Stage breakthroughs will, I believe, yield hitherto-unknown benefits, and possibly exponential leaps in wellness and longevity. At the dawn of the Third Stage, we don't know the answers: researchers are still stumbling merely to formulate the issues and find the right questions to ask. But Olshansky forgot one factor that has unlimited transformational potential, and that is *consciousness.*

Third Stage studies will start from different premises. Why study the effects of infinite potentials on conventional medical treatment, in this case studying the effect of prayer on post-surgical complications? Why not cut out the second stage procedures, and start studying the first causes, the immanent basics of life itself? To reformulate the question in medical terms: *Is there a point at which prayer, hope, and other practices could make obsolete a surgical procedure* like sawing through a patient's sternum in order to do bypass surgery?

8

Belief Therapy

If I told patients to raise their blood levels of immune globulins or killer T-cells, no one would know how. But if I can teach them to love themselves and others fully, the same change happens automatically. The truth is: Love heals.
—Bernie Siegel, M.D., *Love, Medicine and Miracles*

O ne of the most fascinating avenues of Third Stage enquiry is the link between gene expression and belief. Bruce Lipton, Ph.D., is a former professor at Stanford University School of Medicine and an expert on DNA. His best-selling book *The Biology of Belief* has been hailed by Joseph Chilton Pearce as "the definitive summary of the New Biology and all it implies. It synthesizes an encyclopedia of new information into a brilliant yet simple package." His early research on muscular dystrophy, studies employing cloned human stem cells, focused upon the molecular mechanisms controlling cell behavior.

Protean Protein Beliefs

In 1982, Lipton began examining the principles of quantum physics and how they might be integrated into his understanding of the cell's information processing systems. He produced breakthrough studies on the cell membrane, which revealed that this outer layer of the cell was an organic homologue of a computer chip, the cell's equivalent of a brain. His research at Stanford University's School of Medicine, between 1987 and 1992, revealed that *the environment, operating though the membrane,* controlled the behavior and physiology of the cell, turning genes on and off.

Lipton's discoveries, which ran counter to the established scientific view that life is controlled by the genes, presaged the new science of epigenetics. Two major scientific publications derived from these studies defined the molecular pathways connecting the mind and body.[1,2] Many subsequent papers by other researchers have since validated his concepts and ideas[3] Nowadays, Lipton lectures to conventional and complementary medical professionals and lay audiences about leading-edge science and how it dovetails with mind-body medicine and spiritual principles. He has been heartened by anecdotal reports from hundreds of audience members who have improved their spiritual, physical, and mental wellbeing by applying the principles he discusses in his lectures.

Lipton has pioneered the application of the principles of quantum physics—especially the notion that *the quantum universe is a set of probabilities, which are susceptible to the thoughts of the observer*—to the field of cellular biology. While traditional cell biology focuses on the physical molecules that control biology, Lipton's work focuses on the chemical and electromagnetic pathways through which energy in the form of our beliefs can affect our biology, including our genomes. His deep understanding of cell biology highlights the mechanisms by which the mind controls bodily functions, and implies that our bodies can be changed as we change our thinking.

Our beliefs—true or false, positive or negative, creative or destructive—do not simply exist in our minds; they interact with the infinite probabilities of a quantum universe, and they affect the cells of our bodies, contributing to the expression of various genetic potentials. He also shows how even our most firmly held beliefs can be changed, which means that we have the power to reshape our lives.

There are protein molecules on either side of the cell membrane. The proteins on the external surfaces of the cell are receptive to external forces, including the biochemical changes in the body produced by different kinds of thought and emotion. These external receptors in turn affect the internal proteins, altering their molecular angles. The two sets of receptors function like a lattice work that can expand or contract. The degree of expansion determines the size and shape of the molecules—so-called "effector proteins"—that can pass through the lattice. Together the "receptor-effector complex" acts as a molecular switch, accepting signals from the cell's environment that trigger the unwrapping of the protein sleeve around DNA.

Two individuals may have an identical genetic sequence for a particular disease encoded in their cells. The beliefs of the one individual provide the signals that unwrap the protein covering and allow the gene to be activated; the beliefs of the other individual do not. The process is complex; there are hundreds of thousands of such switches in a cell membrane, and which genes are expressed is a function of the configuration of many of them at a given moment. Attempts to correlate a particular belief with the expression of a particular gene cannot be successful because of this complexity. Yet the general principle that our beliefs are affecting the expression of our genes holds true.

Electromagnetic Signaling

As well as chemical signaling mechanisms, Lipton stresses that our bodies use electromagnetic signals to regulate DNA, RNA, and the synthesis of proteins: "In fact, survival is directly related to the speed and efficiency of signal transfer. The speed of electromagnetic energy signals is 186,000 miles per second, while the speed of a diffusible chemical is considerably less than 1 centimeter per second. Energy signals are 100 times more efficient and infinitely faster than physical chemical signaling. What kind of signaling would your trillion-celled community [of cells] prefer? Do the math!"[4] The constantly fluctuating electromagnetic fields of our own hearts and brains are a rich source of environmental information to our cells, as well as signals from other sources. In a truly elegant process, "a tiny field, far too weak to power any cellular activity, triggers a change at the regulatory level, which then leads to a substantial physiological response."[5]

While the work of Lipton and his colleagues shows the precise biological and chemical pathways that influence gene expression, an intriguing series of experiments on the effect on DNA of intention and emotion has been performed by researchers at the Institute of HeartMath in Boulder Creek, California, led by Rollin McCraty, Ph.D. The HeartMath experiments harness the work of Lipton and others to demonstrate practical applications of this knowledge, showing that *measurable molecular changes in the DNA molecule can result from human desires, intentions, and emotions.*

In a series of papers published over the course of the last decade,[6] McCraty and his colleagues have looked at various aspects of heart function and of DNA modulation under different conditions and at various physical distances (to control for the effects of electromagnetic radiation, which can also affect DNA modulation). In one of the Institute's research summaries, they begin by illuminating the relationship of *energetic* to *chemical* transmission mechanisms, a crucial step in understanding the importance of energy in human systems:

"The current scientific conception is that all biological communication occurs at a chemical/molecular level through the action of neurochemicals fitting into specialized receptor sites, much like keys open certain locks. However, in the final analysis, the message is actually transmitted to the interior of the cell by a weak electrical signal. This signal, in turn, can act to either stimulate or suppress enzyme systems. It is now evident that the cell membrane is more than a protective barrier; it also serves as a powerful signal amplifier.

"From these and related findings, a new paradigm of energetic communication occurring within the body at the atomic and quantum levels has emerged—one which is compatible with numerous observed phenomena that could not be adequately explained within the framework of the older chemical/molecular model.

"For instance, our responses to stress have been exquisitely honed over millennia of evolution. 'Fight or flight' reactions to life-threatening situations include shunting of blood away from the gut to the large muscles of the extremities to provide greater strength in combat or speed of locomotion away from a site of potential peril, increased blood flow to the brain to improve decision making, dilation of the pupils to provide better vision, quicker clotting of the blood to reduce loss from lacerations or internal hemorrhage, and a host of other reactions that occur automatically and instantaneously.

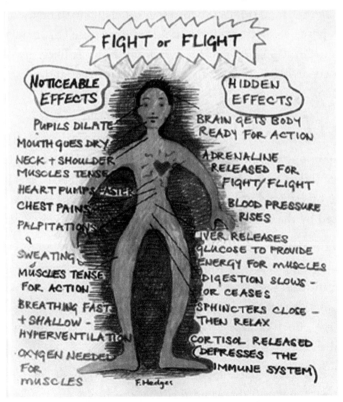

Student drawing of "fight-or-flight" effects[7]

"These responses are too immediate and manifold to be consistent with the key-lock model of communication and thermodynamic laws involving caloric exchange. However, they are comprehensible within the framework of quantum physics and an internal and external electromagnetic or energetic signaling system, which may also explain...the energetic communication links between cells, people, and the environment.

"As we gain new understanding of these fundamental aspects of human function, it will undoubtedly lead to the generation of more effective strategies for improving health, performance, and happiness. We have already found that several of the brain's electrical rhythms, such as the alpha and beta rhythm, are naturally synchronized to the rhythm of the heart and that this heart-brain synchronization significantly increases when an individual is in a physiologically coherent mode. This synchronization is likely to be mediated at least in part by electromagnetic field interactions. This is important as synchronization between the heart and brain is likely involved in the processes that give rise to intuition, creativity, and optimal performance."[8]

DNA Change and Intention

One of the HeartMath experiments used human placental DNA, and determined whether the molecule's helicular coils became more tightly wound or less tightly wound, a characteristic that can be measured by the molecule's absorption of ultraviolet light.

Individuals trained in HeartMath techniques generated feelings of love and appreciation while holding a specific intention to either wind or unwind the DNA in the experimental sample. In some cases, there was a change of 25% in the conformation of the DNA, indicating a large effect. Similar effects occurred whether the intention was to wind the helixes tighter, or to unwind them.

When these participants had no intention of changing the DNA, yet generated the same feelings, the DNA changed no more than it did with the control group, which was composed of local residents and students. When trained participants held the intention of changing the DNA but did not move into the emotional state of love and goodwill characterized by heart coherence, the DNA likewise remained unchanged. Heart coherence is a state in which the variability of our heartbeat is highly regular. Positive emotions produce heart coherence, while negative emotions produce large fluctuations in heart rate variability. A high degree of heart coherence is associated with a high degree of efficiency in the functioning of the circulatory and nervous systems.

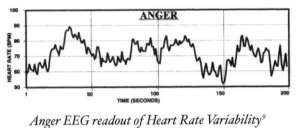

Anger EEG readout of Heart Rate Variability[9]

Appreciation EEG readout of Heart Rate Variability[10]

In order to determine just how specific and local the effect might be, in one experiment with a highly trained volunteer, three separate vials of DNA were prepared. The volunteer was asked to wind the DNA spirals tighter in two of the samples, but not in the third. Those were exactly the results measured under later UV analysis in the laboratory; changes showed up only in the two samples to which the volunteer had directed his intention. This suggests that the effects are not simply an "amorphous energy field," but are highly correlated with the intender's intentions.

The researchers speculated that the effects might be due to the proximity of the samples to the participants' hearts, since the heart generates a strong electromagnetic field. They therefore performed similar experiments at distance of half a mile from the DNA samples. The effects were the same. Five non-local trials showed the same effect, all to statistically significant levels.

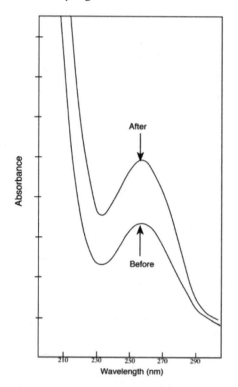

Ultraviolet absorption in DNA before and after being exposed to human intention[11]

These studies demonstrate that the DNA molecule can be altered through intentionality. The better participants were at generating a state of heart coherence, the better they were at affecting DNA with their intentions. Control group participants who were untrained and unskilled at heart coherence were unable to produce an effect despite the strength of their intentions. Both intention and heart coherence were required in order to alter the DNA molecules. "These data," conclude the researchers, "support the hypothesis that an energetic connection exists between structures in the quantum vacuum and corresponding structures on the physical plane, and that this connection can be influenced by human intentionality."

174

The researches used a suspension of DNA molecules in a fluid for the study in order to eliminate any potential biochemical influence that might occur in a live sample. However, they speculate that an individual's own DNA might be more resonant, and therefore more responsive, to that person's intentions. Backing up Lipton's discoveries, they assert that, "the data presented here support the concept that cell-level processes can be influenced by human intention, mediated via energetic interactions."

They also speculate that the positive emotions affecting cell structures might play a role in other phenomena that are well documented but poorly understood. These include the placebo effect, spontaneous remissions, the health and longevity rewards of faith, and the positive effects of prayer.

The Dispensary in Your Brain

Each of us holds the keys to a pharmacy containing a dazzling array of healing compounds: *our own brain*. We are capable of secreting the chemicals that enhance our immune function, make us feel pleasure, and insulate us from pain. For instance, endorphins, the body's natural painkillers, have effects similar to powerful pharmaceutical preparations such as morphine—the word *endorphin* is itself an abbreviation of "endogenous (*internally-generated*) morphine." Consciousness, acting through the body, can generate the molecules required for healing. Our brains are themselves generating drugs similar to those that our doctor is prescribing for us.

Immediate early genes, those that activate in time frames as short as two seconds, are among the genes that stimulate the "fight or flight response" referenced by the HeartMath experimenters. Some of the changes they produce—increased blood circulation to the muscles, enhanced blood clotting, and pupil dilation—are effected by means of proteins. The flight or fight stimulus is particularly interesting because modern humans—we who face few violent perils from the external environment—are capable of turning on these stress responses purely through the quality of our thinking. We can also turn on healing proteins, as researchers studying couple interactions, heart disease, and wound healing have documented.

Research has not yet cataloged all the proteins our bodies produce, but the estimate that scientists use most often is 100,000 proteins. Proteins can be assembled in many different combinations, and these combinations are also the subjects of study. The number of known sequences is at least four million, and rising, according to a team of researchers working on a database of human proteins called the Proteonome.[12] A few of them are household words: insulin, hemoglobin, glutamine, serotonin. They work synergistically with all the other systems in your body.

One of the most intriguing ways in which we activate naturally occurring healing proteins in our bodies is the placebo response. In order for new medications to be proven effective in scientific studies, they are required to demonstrate efficacy that is significantly greater than an inert pill—a placebo. One of the remarkable things about many medications is that their effects are only slightly better than those obtained from a placebo. Some drugs and surgical procedures have results that are no better than a placebo.

A powerful demonstration of the placebo effect comes from Baylor University Medical Center in Texas. The results were published in the *New England Journal of Medicine* in 2002.[13] Bruce Mosely, M.D., an orthopedic surgeon at Baylor, wanted to find out which of two surgical procedures produced the best cure rate for osteoarthritic knees. Many of his patients had degeneration or damage to the cartilage in their knees, and there were two possible surgical procedures that might help.

The first was debridement. In this procedure, incisions are made on both sides of the kneecap and the strands of cartilage are scraped off the surfaces of the knee joint. The second common surgery is called lavage. High-pressure water is injected through the knee joint, flushing out all the old material in the joint.

To control for the placebo effect, Dr. Mosely instituted a third group. This group received neither debridement nor lavage. They patients were prepared for surgery, anesthetized, and wheeled into the operating room. There, the staff made the same skin incisions they would have made were they performing a genuine surgery. After shuffling around the room for the same length of time it took to perform actual surgery, the wounds were sewn up, and the patients released for post-surgical recovery. Absolutely no surgery was done to their knees.

Post-operative results were compared for all three groups during various stages of the recovery process. The results were astounding. Patients that had received the "pretend" surgery did as well as those that had received debridement or lavage. In *Soul Medicine,* I sum up the results this way: "In television interviews, some of the patients who had received the placebo operation swore it must have been real, because their knees were so dramatically improved. Some reported the cessation of pain, or a vastly increased range of motion. Some who were barely able to walk before surgery were now able to run. Not only had the patient group receiving the placebo experienced recovery rates similar to those who had actually received arthroscopic surgery; at certain points in the recovery process their reported results were better. Dr. Moseley, the researchers, and his staff were astounded. There are 650,000 arthroscopic debridement or lavage procedures performed in the U.S. each year. Each one costs some $5,000.[14] The total cost is over three billion dollars per year."[15] These enormous costs appear to accomplish little more than a game of "let's pretend."

It's not just the internal pharmacopia required to cure osteoarthritis that is engaged by the placebo effect. Our bodies are capable of synthesizing many different substances, with radically different chemical structures, within our organs. The four adjoining illustrations of a beta-blocker, histamine, growth hormone, and neuropeptide, represent just four of the millions of compounds your body is capable of synthesizing in response to the placebo effect. The molecules could not look more different, as you can see from their unique shapes. Yet each of these *very different substances* can be synthesized by your body in response to *the same stimulus,* popping a placebo (or fake surgery) to reinforce a belief—a miracle of biological engineering. This ability of our bodies to generate healing factors through belief offers a remarkable therapeutic tool that can be used in conjunction with any therapy, conventional or alternative.

Beta Blocker

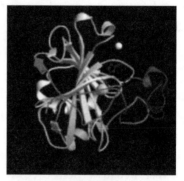

Histamine

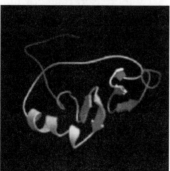

Growth Hormone

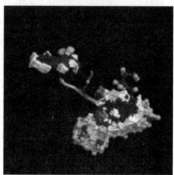

Neuropeptide

Placebos cure patients in about 35% of cases. A drug has to do significantly better for it to be worth taking. Few drugs are up to the challenge. A recent trial of St. John's Wort, an herb, found that 24% of depressed patients got better taking it. Zoloft did marginally better, curing 25% of the patients. But the star of the study was the placebo, which cured 32% of those taking it.[16] In 2006, the U.S. federal government released the results of two large-scale studies of this class of drugs. They found that the tests, "failed to show that the drugs were safer or more effective than a placebo."[17]

Irving Kristol, Ph.D., a psychologist at the University of Connecticut who analyzed the results of drugs studies for depression found that about *three quarters of the entire effect* of antidepressants such as Prozac and Zoloft is due to the placebo effect. Three quarters of the cure was due to the patient's belief system. The remainder may or may not have been due to the drug; it was hard to tell since the drug produces *physical sensations* that may alert study participants to the fact that they're taking a real drug, not the placebo—thus *increasing* the healing factor of the patient's belief system.

Kristol later published a second study, a meta-analysis of forty-seven studies of antidepressants from the FDA database. He found that "an average of 80% of the effect of the drugs was due to the placebo effect. This ranged from a low of 69% (Paxil) to a high of 89% (Prozac). In four of the trials that Kristol studied, the placebo exhibited better results than the drug. The mean difference between the placebos and the drugs was a 'clinically insignificant' difference..."[18]

These results are not atypical; Prozac had to go through *ten* clinical trials in order to accomplish *four* trials with a marginally better cure rate than a placebo.[19] Two trials is what the Food and Drug Administration requires for proof of efficacy; currently drug companies can conduct many trials before they can produce two that show that their drug is marginally more effective than a placebo.

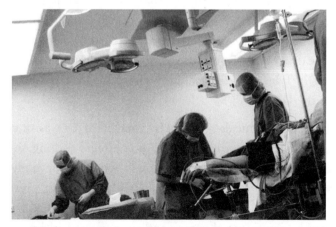

Modern surgery is amazing—yet sometimes unnecessary

The many studies that fail to show that the drug is helpful are called "file drawer studies," because they're quietly filed in dusty corners and never submitted to the FDA or medical journals—though Kristol was able to obtain them to include in his meta-analysis. The result, according to one respected medical journalist, is, "Millions of people taking *drugs that may carry a greater risk than the underlying condition.* The treatment, in fact, may make them sick or even kill them."[20]

While placebos have no side effects, the same cannot be said for prescription drugs. According to the *Journal of the American Medical Association,* some 250,000 people each year die from the negative effects of drugs, unnecessary surgeries, infections they pick up in hospitals, and other iatrogenic (doctor-caused) illnesses, making iatrogenic illness the third largest cause of death in the U.S., just behind cancer and heart disease.[21] According to a meticulous and recent analysis of data from a spectrum of U.S. government agencies, when all other factors are included, the number rises to 783,966, *making doctors, hospitals, and drugs the leading cause of death in America.* Each year, this costs society an estimated $282 billion. About *twice* the number of Americans die *due to infections acquired in hospitals* each year as the number who die in traffic accidents. And each year, more than *twice* as many Americans die from *the negative effects of prescription drugs* as those who died in the Vietnam war. According to a recent report, an estimated 1,730,000 preventable drug-related injuries occur in America each year, and as many as 7,000 deaths.[22] And not only exotic drugs are the culprits; common acetaminophen, the active ingredient in Tylenol, is the leading cause of death due to acute liver failure in the U.S.[23]

The neurochemicals secreted by your brain, by way of contrast, do not come packaged with a list of side effects so long, extensive, and severe that the average patient will wince. Your body's natural secretions are often the same substances found in prescription drugs—but in doses that will not harm you, in forms that are readily assimilated by the targeted organs and systems, and without the side effects of prescription medications. You prescribe them for yourself by the quality of your consciousness.

One study, at the University of Madison, Wisconsin, recruited thirty-six Vietnam veterans with coronary artery disease, and who were also plagued with traumatic emotional issues. Half the vets received training in how to forgive themselves and others—a consciousness-based medical intervention—and the other half did not. Those that had received the training and practiced the forgiveness techniques showed a significant rise in the blood flow to their heart muscles.[24] Such consciousness-based interventions are safe and noninvasive, and are a good place to begin treatment. At the very least, they can supplement the arsenal of allopathic medicine; in the many cases of "miraculous" cures, they make medical treatment unnecessary.

Quantum Puzzles

To get the right answers, you have to ask the right questions. To the scientists of a generation ago, the idea that DNA might be malleable was unthinkable. Genes were taken to determine the physical characteristics of organisms, and that was that. Researchers didn't look for answers that might suppose pliable DNA, or combinations of proteins variable by thought, because they were embedded in an orthodoxy that held genes to contain the DNA sequencing around which physical, mental, and emotional development occurred.

It took a bold new generation of experimenters to ask the right questions, questions that wondered whether DNA could change as environmental factors changed, and whether our body's own pharmacy might supply the chemistry needed for optimal wellbeing—as well as or better than the prescription pad of the physician. Once they had been asked, experiments could be set up that would yield fresh new answers, answers that flew in the face of orthodoxy but held keys to a scientific explanation of such phenomena as the effectiveness of energy medicine. Today, an increasing body of experiments proceeds along this line of inquiry. As the right question tickles the curiosity of more researchers, experiments—and answers—will multiply.

181

9

Entangled Strings

Only in the last moment in history has the delusion arisen that people can flourish apart from the rest of the living world.

—E. O. Wilson

"In the early 1990s I was in Toronto, Canada. I went to see my doctor because I felt tired and listless. He sent me to have an electrocardiogram. Later that day, when he got the results back, he told me that my heart was at serious risk. He told me to stay calm, not exert myself, keep nitroglycerine pills with me at all times, and to not go outside alone.

"The doctors administered several tests over the course of the following three days, and I failed them all because my arteries were severely clogged. They included an angiogram, another electrocardiogram, and a treadmill stress test. When I started the bicycle test, the clinic staff didn't even let me finish. They stopped me part way. They were afraid I was going to die on the spot, my arteries were so clogged. As a high-risk patient, I was given an immediate appointment for heart bypass surgery.

183

"The day before the surgery, I woke up feeling much better. I went to the hospital and I was given an angiogram. This involved shooting dye into my arteries through an injection in my thigh. The surgeons wanted to discover the exact location of the blockages prior to the operation. I was prepared for surgery. My chest was shaved, and the doctors were about to mark my skin where they planned to make the incision. When the new angiograms came back from the lab, the doctor in charge looked at them. He became very upset. He said he had wasted his time. There were no blockages visible at all. He said he wished his own arteries looked as clear. He could not explain why all the other tests had shown such severe problems.

"I later discovered that my friend Lorin Smith in California, [a Pomo Indian medicine man] upon hearing of my heart trouble, had assembled a group of his students for a healing ceremony the day before the second angiogram. He covered one man with bay leaves and told him that his name was Richard Geggie. For the next hour, Lorin led the group in songs, prayers, and movement. The next day, I was healed.

"I have seen Lorin facilitate other amazing healings. Sometimes he works in a trance, invoking his grandfather, Tom Smith, who was a very famous healer. When he emerges from trance, he's unable to remember what he has said."

This story was told by former cardiac patient Richard Geggie for the book *The Heart of Healing*. Some thirteen years later, Geggie was still in excellent health.[1] It is by no means an isolated example of distant healing; there are impressive databases dedicated to cataloging studies of the phenomenon.[2]

Two careful studies of terminally ill AIDS patients by Elizabeth Targ, M.D., Ph.D., showed similar results. The patients in her experimental groups were offered remote healing by forty religious and spiritual healers. Some were evangelical Christians, some were traditional Catholics, some were Buddhists, some were independent faith healers; one was a Jewish kabbalist, another was a Lakota Sioux shaman, and another was a Chinese Qigong master. After going through several steps to ensure that no one in the study could know which patients were being prayed for, the healers were sent an envelope containing the patient's photo, name, and helper T-cell count.

Six months later, the researchers found that those who had received remote healing showed improved mood, significantly fewer doctor visits, fewer hospitalizations, fewer days in hospital, improved helper T-cell counts, fewer new AIDS-defining illnesses, and significantly lower quantities of the HIV virus in their systems.[3]

The results of this study of healing at a distance are not unusual. Researchers have conducted several meta-analyses of distant healing research, and most of these have found results far in excess of what might be explained by chance.[4][5][6]

Electromagnetic signaling explains many healing phenomena that cannot be explained by other medical models. The semiconductive properties of connective tissue explain many others. However, healing at a distance, or nonlocal healing, cannot be explained by either mechanism. How is it possible for healing to occur when the person receiving the healing is too far away from the person offering the healing to be affected by the healer's electromagnetic fields? Quantum physics and string theory offer us some intriguing insights as to how this might occur.

Subatomic Alchemy

In 1907, a pioneering physicist, Ernest Rutherford of the University of Manchester, discovered that the atom could be subdivided into a nucleus of protons and neutrons, surrounded by an orbit of electrons. He showed that virtually the entire volume of an atom was empty space, rather than the solid substance conceived of in Newtonian physics, and that most of the atom's mass was concentrated in the nucleus. He also demonstrated that electrons were bound to the nucleus by electromagnetic fields. So enamored was he of his profession that he declared: "In science there is only physics; all the rest is stamp collecting." For the ensuing century, medicine returned the compliment by ignoring the discoveries and implications of quantum theory for healing.

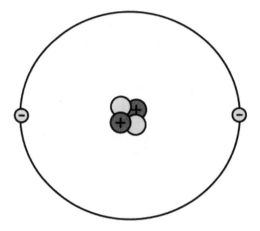

Atomic model circa 1907:
Particles of a helium atom—two electrons orbiting
a nucleus of protons and neutrons

By the start of the twenty-first century, physicists had cataloged dozens of subatomic particles—among them baryons, pentaquarks, mesons, leptons, axions, neutrinos, muons, and bosons. Many particles have characteristics that challenge our perception of linear time and orderly space, by manifesting into existence, then disappearing again only to repeat the cycle. We have observed matter vanishing into energy, then winking back in somewhere else as a different type of particle.

Even more unsettling, rather than fixed realities, these energies exist in the form of an infinite set of possibilities, a "possibility wave," out of which probabilities emerge. Quantum physics can calculate a range of "possible events for electrons and the possibility of these possible events, but cannot predict the unique actual event that a particular measurement will precipitate."[7]

Theoretically, any one of the swarm of infinite possibilities present in the possibility wave can become reality. But only one does. The swarm is then said to have "collapsed" into a particular reality. One of the factors that determine the direction in which the swarm of possibilities collapses is the *act of observation*. In a quantum universe, phenomena and space and time are affected by the observer. All possibilities exist in the quantum field; the act of observation collapses them into probability.

"In the realm of possibility," says quantum physicist Amit Goswami, Ph.D., "the electron is not separate from us, from consciousness. It is a possibility of consciousness itself, a material possibility. When consciousness collapses the possibility wave by choosing one of the electron's possible facets, that facet becomes actuality."[8] So the scientific mind, rather than impartially witnessing objective phenomena, is itself influencing which of the infinite sea of potentials winks into existence as a phenomenon. Goswami continues, "The agency that transforms possibility into actuality is consciousness. It is a fact that whenever we observe an object, we see a unique actuality, not the entire spectrum of possibilities. Thus, conscious observation is sufficient condition for the collapse of the possibility wave."[9]

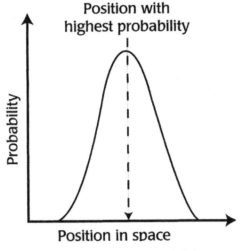

Distribution of quantum probability[10]

The Observer Effect was first observed in physics, though it holds true in other disciplines. Douglas Yeaman and Noel McInnis remind us in the book *Einstein's Business* that, "A major scientific precedent for managing our outcomes was set by quantum physicists who, when seeking to determine whether light consists of particles or waves, discovered that light invariably behaves in compliance with their experimental expectations. Light always and only behaves like waves in experiments designed to detect waves, yet just as consistently shows up as particles in experiments designed to detect particles. In both cases, experimental outcomes conform to the experimenters' expectations." Physicist Werner Heisenberg said, "What we observe is not nature itself, but nature exposed to our method of questioning." In *The Intention Experiment,* her book about a large international experiment underway to gauge the effect of human intention on physical matter, Lynne McTaggart says that the observer effect implies that, "living consciousness is somehow central to this process of transforming the unconstructed quantum world into something resembling everyday reality," and that, "reality is not fixed, but fluid, and hence possibly open to influence."[11]

A gifted faith healer might be considered, in quantum terms, to be an observer who routinely collapses space-time possibilities into the probability of healing. A prayer is an intention that might also collapse the swarm of possibilities present in the possibility wave in the direction of a certain probability. Such quantum phenomena are not limited by geography or history. Einstein said, "The distinction between the past, present, and future is only a stubbornly persistent illusion."

Hundreds of experiments have shown the effect of an observer on material reality. A particularly charming and visual demonstration of the effect of intent was done by Rene Peoc'h, M.D. He used a self-propelled robot he called a "Tychoscope," which contained a random number generator. He monitored its movements on a plotter. When it was put in a room, the plotter tracking the robot's movements found that it indeed moved around the room in an entirely arbitrary way (A), at random angles and for random lengths of travel.

He then introduced a cage of chicks at one end of the room. The chicks had been trained to relate to the Tychoscope as though it were the mother hen. In the presence of the chicks' intention, the robot's movements were no longer random; it approached the cage of chicks two and a half times more often than it would approach an empty cage (B).[12]

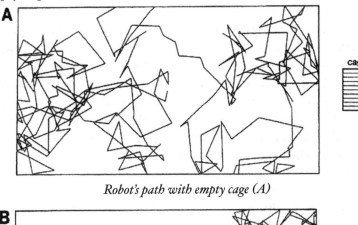

Robot's path with empty cage (A)

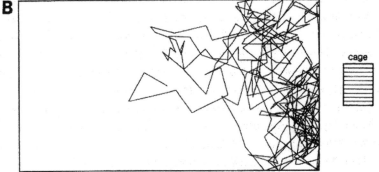

Robot's path with with chicks (B)

Another possible explanation for distant healing comes from string theory. String theory explains phenomena like electron tunneling. "Tunneling" is a misleading term, since it implies that electrons travel through a tunnel. But in fact, they don't. Electrons in a field have been observed to jump from one orbit to another orbit without traveling through the intervening space, and without any time having elapsed. They disappear in one orbit, and simultaneously reappear in another orbit. Physicist Niels Bohr was the first to describe such a "quantum leap."[13] One researcher describes tunneling this way: "Suppose that you fire a particle such as a proton or an electron at some kind of wall that it doesn't have enough energy to penetrate.... Occasionally the particle will appear to tunnel straight through what would otherwise be an impassable obstacle, just by happening to jump from one part of its probability wave to another."[14]

How can an electron jump from one spot to another without going through the space between the two points, and without taking any time to complete the process? Even Albert Einstein was baffled by the phenomenon.[15]

String theory solves the problem by postulating an eleven-dimensional multiverse. Electrons may jump out of existence in our universe, into another, and then reappear in our universe. The tunnel through which the electrons pass exists in another universe within the multiverse, and is not subject to the rules of space and time that govern the limited number of dimensions of which we are aware.

String theory also explains how subatomic particles can demonstrate both the properties of a particle and of a wave. We are accustomed to thinking of things as either a particle (a solid object), or a wave (a vibration). Yet light, and many subatomic particles, can effortlessly manifest as either a wave or a particle.

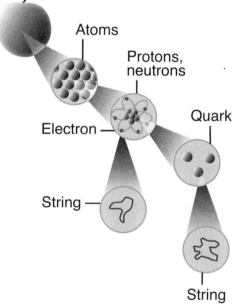

Atomic model circa 2007:
Vibrating at different frequencies, strings create matter[16]

According to string theory, everything in the universe is made up of tiny vibrating strings. These strings are identical. The reason that one manifests as a heavy particle, like a proton, and another as a particle without mass, like light, is that they vibrate at different frequencies. According to Einstein's famous equation $E=mc^2$, the more *energy* something has, the more *mass* it has. So a heavy particle is a string vibrating at a high frequency, while a light particle ("light" in both senses of the word) is a string vibrating at a lower frequency. Strings are also very small; if the image of an atom illustrated at the start of this section was magnified to the size of our solar system, a string would be the size of a tree.

The eleventh dimension is the dimension that may contain within it, in the words of one BBC reporter: "any number of parallel universes. Physicists say that this dimension may be only a millimeter away from us yet we have no awareness of its existence."[17] This is because the other universes are vibrating *out of phase* with ours, at a frequency that we cannot perceive.

There is vastly more mass in our universe than we can perceive. This "missing mass"—some 96% of the total—is thought to consist of "dark energy" and "dark matter," terms which, according to *Scientific American,* "serve mainly as expressions of our ignorance, much as the marking of early maps with 'Terra Incognita.'"[18] This dark matter and dark energy may be manifestations of the parallel universes.

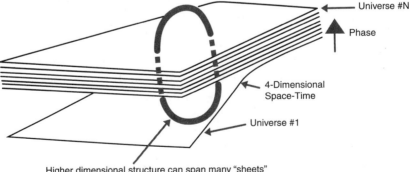

Universe #N

Phase

4-Dimensional Space-Time

Universe #1

Higher dimensional structure can span many "sheets" corresponding to stable universes. These can bring knowledge or energy from other dimensions.

Model of a multi-dimensional universe[19]

The alternate universes postulated by string theory are like pieces of paper stacked one on top of the other. Each vibrates at a slightly different rate, so we cannot perceive them. As the vibrational rate of a string changes, it may jump out of phase in our universe, into phase in another, and then back into phase in ours. As the string resonates at a particular frequency, it produces subatomic particles. Each resonance is associated with a particular particle. As the vibrational frequency of the string changes, it expresses as different subatomic particles. Like the tunneling electron, these particles can flash into existence in our universe, then flash into another universe, and then flash back into ours. In this way, signals can pass between dimensions.

Discomfort and Novelty in Research

When looking for explanations of how the world works, we are historically much more comfortable with solid objects (particles)—like the ion transfer theory of classical neurology, or the use of drugs and chemotherapy to treat cancer. We are less comfortable with intangible phenomena (waves)—like electromagnetic signaling, or the effects of consciousness and energy healing on cancer. String theory rattles our cage, stretches our minds, and reminds us that healing can occur in dimensions that are outside of our known universe, let alone our comfort zone.

Physicist Michio Kaku, one of the originators of string theory, compares us to a fish, swimming with other fishes in a shallow pond. This fish has explored every facet of the two-dimensional universe that it inhabits. Then someone yanks the fish from the pond, and lifts it high above the pond. Suddenly the fish experiences a third dimension, "up," of which it was previously unaware. It comes back to the pond and tries to explain the remarkable perspective from which it has just seen the world. The other fish are baffled.[20]

That state of bafflement is common when we attempt to stretch our space-time-delimited minds around the complexities of quantum physics and string theory. Yet they may explain phenomena such as distant healing, which we know occur, yet which current medical models, bound as they are to the pond in which we swim, cannot explain or comprehend even with the aid of electromagnetism and rapid connective tissue signaling.

While you may be sick in this dimension, you may be the model of health in another dimension of the multiverse. Your intention in this dimension, acting in a quantum field, may collapse the swarm of probabilities in such a way as to influence health. Such ideas stretch the boundaries of our imagination, yet they hark back to shamanic journeys, or aboriginal walkabouts in the dreamtime, into which our ancestors traveled to find cures for what ailed them.

An Entangled Universe

Another characteristic of quantum theory is entanglement—the idea that relationship is the defining characteristic of everything in space and time. The concept of entanglement arose from the observation by physicists that certain particles still appear to move in a connected fashion, without any time lag, even after they are separated by large amounts of space. In his book *Entangled Minds,* researcher Dean Radin quotes Erwin Schrödinger, one of the early proponents of quantum theory, and the originator of the name "entanglement," as saying: "I would not call [entanglement] *one* but rather *the* characteristic trait of quantum mechanics."[21] Not only does atomic matter become entangled; whole systems may become entangled too. Radin quotes a March, 2004, review of the field of entanglement by *New Scientist* magazine which concludes that: "Physicists now believe that entanglement between particles exists everywhere, all the time, and have recently found shocking evidence that it affects the wider, 'macroscopic' world that we inhabit."[22]

Given the pervasive nature of the entanglement of atomic particles, it is unlikely that evolution has ignored this basic characteristic of matter in setting up human systems. Having realized that some form of epigenetic control is required to produce the complex orchestration of living systems, entanglement is a good place to look for the mechanisms of cybernetic control. According to a researcher at the

Vienna University of Technology, "Entanglement could coordinate biochemical reactions in different parts of a cell, or in different parts of an organ. It could allow correlated firings of distant neurons. And... it could coordinate the behavior of members of a species, because it is independent of distance and requires no physical link."[23] British physicist David Boehm observed that entanglement applies "even more to consciousness, with its constant flow of evanescent thoughts, feelings, urges, desires, and impulses. All of these flow into and out of each other."[24] Thus entanglement offers a tantalizing glimpse into how distant intention and intercessory prayer might operate to produce healing even at great distances.

Dr. Robin Kelly, a British physician, has accumulated much of the research on the properties of microtubules. Since they are resonant structures, he asks, what is the signal with which they may be resonating? He speculates that they receive signals from the quantum field, and play a role in the communication between one cell and another, and between all cells and the quantum field. Kelly describes how, "In the early 1990s a British physicist, Sir Roger Penrose, joined forces with an American anaesthetist, Dr. Stuart Hameroff, as both were intrigued by these microtubules. Sir Roger was naturally drawn to the spiral structure and wondered if he could be looking at a sophisticated quantum computer; a processor of information gleaned from the quantum world. To Dr. Hameroff, the microtubules resembled a computer switch." They "developed the hypothesis that these hollow tubes were our body's link with consciousness—an environment where the timeless quantum world was allowed to collapse down to our recognizable physical world of time and space."[25] While the resonant properties of microtubules is a subject still awaiting serious research scrutiny, it is possible that they may play a role in the transmission of intention and consciousness across distances.

The Mirror in Your Brain

One of the most exciting new areas of brain research currently is the field of mirror neurons. Mirror neurons are neurons that fire in our brains when we witness an act done by another that requires the same group of neurons. When the neurons in their brain fire and they perform the action, the neurons in our brain fire in sympathy. When another person reaches for a glass of water, for instance, the neurons in our brains that govern that action fire when we witness their action, just as the other person's neurons do when they perform it. This whole process happens without going through the normal sensory-cognitive cycle that conventional brain models would suggest: visual images passing from the optic nerve (as our eyes witness the other person performing the action), then to the visual cortex, then to the parts of the brain that govern decision-making, and eventually being translated into signals sent via the nervous system.

The first researchers to describe the phenomenon were a team of Italian scientists at the University of Parma led by Giacomo Rizzolatti. They were studying the brain patterns of macaque monkeys, in particular the pre-motor cortex, which plans and initiates movements, which then travel to the muscles via the nerves.

The researchers identified which brain cells became active when a monkey picked up a nut. However, one of the researchers inadvertently picked up his own nut while a monkey was still attached to the brain scanner. Much to the surprise of the scientists, the same area of the brain became active in the monkey while observing the researcher's action.

This initial discovery, in 1995, led the team to find many other groups of neurons that mirror the actions of others.[26] Other researchers in other countries are now investigating the properties of mirror neurons.

An early clinical application was the study of autism. Researchers at the University of San Diego, led by Vilayanur Ramachandran, suspected that autistic children might not be picking up the same cues from the actions of others. When they compared a group of autistic children with a control group, they discovered that indeed, the autistic group did not register mirror neuron activity.[27]

Albert Einstein and other scientists first proposed, in 1935, that nonlocal action was a requirement of quantum mechanics (the Einstein-Podalsky-Rosen Paradox[28]), but it was not verified experimentally until much later. To find out if quantum entanglement also occurs between human beings, a group of experimenters designed and conducted an innovative study. They arranged for two people to meditate, sitting side by side, in a Faraday cage—a room specially constructed to screen out all electromagnetic and other radiation. After a time, one of the subjects was moved to a different Faraday cage about 15 yards away, and hooked up to an EEG machine. Then a light was shone, at irregular intervals, into the eyes of the meditator in the first room, who was also hooked up to an EEG.

The EEG recordings were later compared. The periods in which the brain of the person into whose eyes the light was shone recorded changes in their EEG pattern were noted and compared to the EEG readouts from the person who had been moved to the second Faraday cage, as well as to the readouts for the control group. The researchers found that at the precise moment when the light was shone into one meditator's eye, the brain of the other meditator responded, while the brains of the control group demonstrated no such effect. The concluded that a human brain is capable of establishing nonlocal relationships with other brains, even when sensory communication, electromagnetic signals, and other cues have been ruled out.[29]

While scientists are eagerly putting such observations to practical use, it's worth pausing to measure the speed at which these mirror effects occur. We've assumed that mirror neurons fire when an action is visually observed, but the firing may in fact occur faster than the speed of visual observation. What if we discover that mirror neurons fire simultaneously in the observer and the observed? Such linkages might be routine in an entangled universe in which particles change orbits without any intervening time. The shared activation of mirror neurons may be a quantum field effect, the physical manifestation of the quantum links that bind us together in ways we scarcely suspect as we go about our everyday lives. Brain cells also contain very high concentrations of microtubules, and these resonant structures might eventually be shown to play a role in the simultaneous transmission and reception of information. Synchronicity has been studied by scientists for a century, and continues to yield fresh insights, such as this story recounted by Marie-Louise von Franz:

> Darwin had developed his theory in a lengthy essay, and in 1844 was busy expanding this into a major treatise when he received a manuscript from a young biologist unknown to him. The man was A.R. Wallace, whose manuscript was a shorter but otherwise parallel exposition of Darwin's theory. At the time, Wallace was in the Molucca Islands of the Malay Archipelago. He knew of Darwin as a naturalist, but had not the slightest idea of the kind of theoretical work on which Darwin was at the time engaged. In each case, a creative scientist had independently arrived at a hypothesis that was to change the entire development of biological science. Backed up later by documentary evidence, each had initially conceived of the hypothesis in an intuitive "flash."[30]

The number of joint Nobel prizes, often awarded to researchers who have had no contact with each other, testifies that entanglement did not start or end in the nineteenth century. Near the beginning of the twentieth century, a British mathematician and physicist, Sir James Jeans, presciently observed: "When we view our selves in space and time, our consciousnesses are obviously the separate individuals of a particle-picture, but when we pass beyond space and time, they may perhaps form ingredients of a single continuous stream of life. As it is with light and electricity, so it may be with life: the phenomena may be individuals carrying on separate existences in space and time, while in the deeper reality beyond space and time we may all be members of one body."[31]

10

Scanning the Future

Research is the highest form of adoration.
—Pierre Teilhard de Chardin[1]

One evening, Albert Einstein's son-in-law, Dmitri Marianoff, sat with him in a house in Berlin, Germany, after all the other members of the family had gone to bed. Into the pregnant stillness, Marianoff asked a question that had long intrigued him:

"'How is it, Albert, that you arrived at your theory?'

"'In a vision,' he answered.

"He said that one night he had gone to bed with a discouragement of such black depths that no argument would pierce it. 'When one's thought falls into despair, nothing serves him any longer, not his hours of work, not his past successes—nothing. All reassurance is gone. It is finished, I told myself, it is useless. There are no results. I must give it up.'

"Then this happened. With infinite precision the universe, with its underlying unity of size, structure, distance, time, space, slowly fell piece by piece, like a monolithic picture puzzle, into place in Albert Einstein's mind. Suddenly clear, like a giant die that made an indelible impress, a huge map of the universe outlined itself in one clarified vision.

"And that is when peace came, and that is when conviction came, and with these things came an almighty calm that nothing could ever shake again..."[2]

The creative flowering of consciousness is as mysterious as Einstein's vision. After he had that insight, it then took him another four years to work out his seminal equations showing the link between energy and matter. But the first impulse was a gift from the universe, seeding a mind open to receiving a new way of seeing the cosmos.

Experiments that measure the interaction of our consciousness with matter hold many surprises. They show us that many of the linear, cause-and-effect relationships that underpin our perception are inventions of our brains, and not the way the world actually works.

Interactive Fields

Researchers at the Institute of HeartMath have done a series of experiments on the effects of consciousness on cells. These experiments are done with rigorous protocols, and are intended to replicate earlier research.[3] They extended the work of Dean Radin, Ph.D., Senior Scientist at the Institute of Noetic Sciences in Petaluma, California. Radin and some of his colleagues measured the galvanic skin response (electrodermal activity) of subjects exposed to the mental influences of others.[4] In a follow-up study, which replicated the results of earlier studies done by them and others, the researchers set up sixteen sessions. In each session, there were seven people acting as mental influencers, and ten acting as remote targets of influence. Influencers were instructed to either calm or activate a remote person's electrodermal activity.

The investigators found that, to a statistically significant degree, when the influencers attempted to calm the subjects, the subjects exhibited a lower level of electrodermal activity. When the influencers attempted to activate the subjects, the subjects showed a higher level of electrodermal activity. Building on Radin's research, the experimenters at HeartMath went further. Rather than the relatively simple and uninformative measurement of galvanic skin response, they also used an electroencephalogram (EEG) in order to measure changes in the cerebral cortex, and an electrocardiogram (ECG), to measure the acceleration or deceleration of a subject's heartbeat.

Rather than a remote influencer attempting to influence their experiences as Radin had done, with all the uncertainties inherent in human processes, the HeartMath subjects stared at a blank white computer monitor screen. After a period of a few seconds, an image came up on the screen. One set of images was designed to calm the subjects, as measured by brain and heart responses, and the other set of images was designed to produce emotional arousal. The images were generated at random by the computer just before the instant of projection from amongst forty-five images stored on the hard drive.

The researchers wanted to find out precisely where and when emotional arousal occurred in the body, heart, and brain. They also presented the images to the subjects under two sets of experimental conditions. One was a baseline condition of normal physiological function. The second was a state of heightened heart coherence, in which their hearts were beating at an unusually even rate.

They discovered that the heart responded to the images. This was not surprising. What was surprising was that it responded *first,* before any mental activity had shown up on the EEG. It appears that the heart communicates its perceptions to the brain, rather than vice versa. But the truly astonishing finding of these experiments was that both heart and brain responded *before* the image had flashed onto the screen—before the random image generator in the computer had generated any image at all. Heart and then brain responded to the type of image *about to be* flashed on the screen—several moments before the computer made its random choice and presented it to the subject. The subject's body then responded appropriately to the emotional stimulus of the image, even though in the objective real world, that stimulus had not yet been presented to either heart or brain. In the words of the amazed researchers: "This study presents compelling evidence that the body's perceptual apparatus is continuously scanning the future."[5]

Canadian researchers, in a series of experiments published as early as 1949, noted an associated phenomenon. The subjects were epileptics undergoing brain surgery, and electrodes were placed directly on the cerebral cortex. The subjects were told, at intervals, to move their fingers. Cortical scans showed that some of the subjects recorded an increase in activity just before the instruction, "Get ready to move your fingers," was given by the researchers. Another research team replicated these results in 2000.[6] In a Newtonian universe that knows only linear time, such a phenomenon is scientifically impossible. Only a quantum universe of fields that interact continuously through time and space can explain such phenomena.

Another example of the power of prayer across time comes from a study published in the *British Medical Journal* in 2001. In Israel, Professor Leonard Leibovici took a stack of hospital case histories and divided it into two random piles. The patients in these cases had all been admitted for blood poisoning. Names in one stack were prayed for, while the others were not.

On later analysis, the group prayed for was found to have a reduced rate of fevers, shorter hospital stays, and a lower mortality rate. This kind of finding is typical of prayer studies and would not have surprised most researchers—except that the patients Leibovici prayed for *had been discharged from hospital ten years earlier.* The healing power of consciousness and intention appears to be independent of time as well as space.[7] Prayer seems to work retroactively as well as across great distances. Perhaps the whimsical injunction is true: "It's never too late to have a happy childhood!"

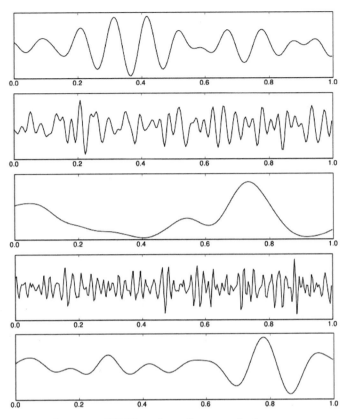

One-second EEG readout of common brain waves.
From top: alpha, beta, delta, gamma, theta

Non-Local Perception

A second study by the HeartMath researchers examined where and when in the body, heart, and brain intuitive information outside the range of conscious awareness is processed. They found that the primary areas of the brain involved are the frontal cortex, temporal, occipital, and parietal areas, and that these are all influenced by the heart. They concluded that, "Our data suggest that the heart and brain, together, are involved in receiving, processing, and decoding intuitive information. On the basis of these results and those of other research, it would thus appear that intuitive perception is a system-wide process in which both the heart and brain (and possibly other bodily systems) play a critical role."[8]

203

"The heart has access to realms of quantum information not constrained by time and space," Dr. McCraty told me during a telephone interview, many years after I first met him while working on some of his institute's early publications. He continued, "There is no explanation other than that consciousness is non-local and non-temporal."

McCraty is preparing a new set of papers, to postulate a theory based on holographic principles that explain how intuitive perception allows us to gain access to an energy field that contains information about "future" events. He is also preparing a rigorous new set of protocols for experiments that will use live cells from the subject's own body to see if there is a similar prior effect in those cells to intentions generated remotely by the subject.

John Arden, Ph.D., chief psychologist at the Kaiser Permanente Medical Center in Vallejo, California, presents a long and careful discussion of theoretical physics, subatomic particles, and their implications for the study of consciousness, in his book *Science, Theology and Consciousness*. He concludes by reminding us that, "nonlocality is a phenomenon operative in nature. This discovery necessitates a fundamental reevaluation of causality and the nature of nonlocal interaction."[9] In other words, our mechanistic notions of cause and effect in time and space operate on a very limited range of the spectrum of what is possible. Though we see ordinary examples of distributed nonlocal consciousness every day, such as schools of fish that turn in tandem, or flocks of birds that bank and swoop in perfect coordination, our collective medical brain still has trouble with the idea that things far apart in space and time can affect each other.

The Half-Second Delay

Benjamin Libet, Ph.D., conducted a provocative set of experiments in which he noted the precise instant at which brain activity indicated awareness of a sensation on the skin. He measured when the skin became aware of the sensation, and when the brain did. This led to the discovery of the "half-second delay." Libet's experiments measured the difference in time between our performance of a muscular action, such as reaching our arm out to grasp an object, and our conscious awareness of the action—the firing of the part of our brain that corresponds to that particular type of muscular movement, and the point in time where we might say, "I move my arm."

Libet discovered that our consciousness projects itself backward in time, to believe that it became conscious of a stimulus about half a second before it actually did so. The brain is convinced that it became aware of an action before it occurred, while in reality it became aware of the action a half-second later. In his funny, provocative, and oft-quoted book *The User Illusion: Cutting Consciousness Down To Size,* Danish science writer Tor Norretranders, explaining Libet's work, says, "The show starts before we decide it should! An act is initiated before we decide to perform it!"[10] He goes on to say, "Man is not primarily conscious. Man is primarily nonconscious. The idea of a conscious 'I' as housekeeper of everything that comes in and goes out of one is an illusion; perhaps a useful one, but still an illusion."[11]

Brad Blanton, Ph.D., in his book *Radical Honesty*, applies Libet's work to practical psychotherapy.[12] He points out that the half-second delay means that the mind thinks up rationalizations to explain what we did—after we already decided to do it. While we present these as reasons for our actions, Libet's experiments show that these rationalizations occur after the fact. Blanton's therapy is aimed at bringing us as close to the moment of action as possible and abandoning our rationalizations for that moment—along with the layers of interpretation and story we build upon our foundation of rationalizations. Physicist Roger Penrose summarizes Libet's results in his book *The Emperor's New Mind,* and then speculates that, "I suggest that we may actually be going badly wrong when we apply the usual physical rules for time when we consider consciousness!"[13] Science is catching up to where spirituality has been for thousands of years; the Buddha urged us to maintain a calm state of desireless attention to the present moment, reminding us that, "We are what we think. All that we are arises with our thoughts. With our thoughts, we make the world."

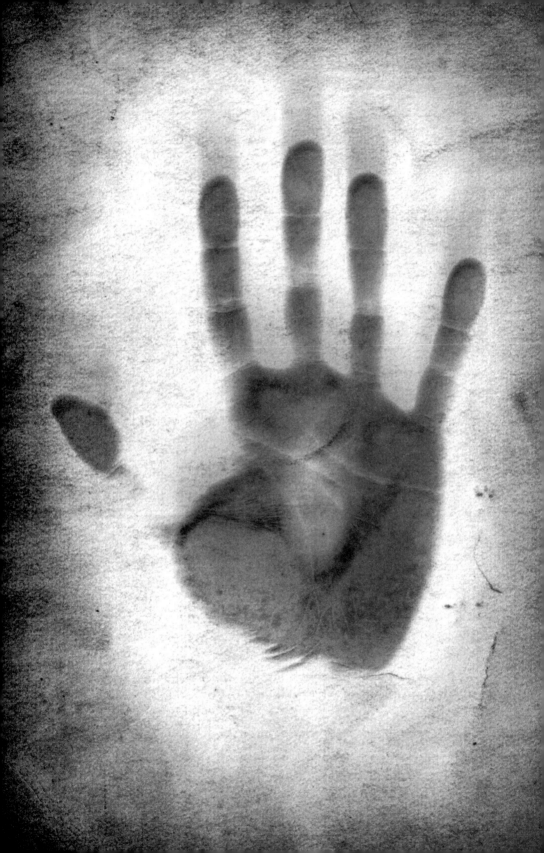

11

Routine Miracles

Based on what we know of the plasticity of the brain, we can think of things like happiness and compassion as skills that are no different from learning to play a musical instrument or tennis...it is possible to train our brains to be happy.

—Neuroscientist Richard Davidson, Ph.D.

"Irv was just 54 years old. He had been a cop, and then had worked as a private investigator for about 25 years. He liked to go sailing in Sheepshead Bay, and did woodworking in his garage. When the heart attack hit, it felt like someone had slugged him in the chest with a 4x4 beam. He says he remembers just dropping like a leaf in the wind.

"His wife and son held his hand at the bedside. The Columbia Presbyterian heart surgeon pursed his lips and looked down. 'He's been in a coma for five days in spite of everything we've done. He's on the transplant list, but I don't think he'll make it long enough for us to find him a new heart.' They cried and kissed him, and said goodbye.

"It was five months later when Irv walked into my office. He was still a bit shaky, way overweight, and he looked pale. But he sat down, smiled, and said, 'I shouldn't be here, Doc.'

"'How come?' I asked. He looked down at his hands. 'Well, because I died five months ago...'

"Or so it seemed to him. A heroic last-option quadruple bypass surgery saved his life—but just barely.

"Irv had pretty bad diabetes for about fifteen years. That set him up for his severe heart attack, and a lot of other problems. It also left him with numb and painfully burning feet and hands. I couldn't do much about his diabetes; that was a job for another doctor, but as a pain doc I sure could help him with the burning damage to his nerves.

"I adjusted his nerve medications, put him on a couple of new things, a couple of supplements, and then we just sat and talked for about twenty minutes. This was a man who had gone through a huge life crisis, and was actually still in crisis, still very wobbly. His struggle was something palpable in the room, like the shadow of death, still there. He talked about his family, his job, what had been important to him, his past sense of meaning and religious practice, and the future. Then he started to cry.

"We talked for a while, then I looked down at my prescription pad and I began to write. As a Scottish Episcopalian who was not particularly religious, I had never written such a thing before in my career. I handed it to him and he looked down at the script. The prescription read 'Long conversations with your Rabbi, twice a week.' He left the office with an odd smile on his face.

"Over the subsequent six months, Irv got more involved with his synagogue. He took the opportunity to allow his terrifying brush with death to work a deep magic on his sense of himself and his day-to-day life. He finally found himself much less concerned with the small things. He still had significant symptoms, but the residual burning in his feet and hands did not bother him as much. He spent more time with his son. He complained much less and got out of the house much more.

"He came back to me and told me he had started a program, through his temple, to provide services for older people in the community. His years as a private investigator had given him the ability to find many resources for his clients, and to protect them from fraud. He said that this was what he was meant to do, and that he could not have found it without going through his illness. He allowed the pain and suffering to become his greatest teacher. As he talked to me about his new calling, his face was bright and his hands were steady. The shadow of death was gone from the room."[1]

Prescription: Belief

Imagine the healing arts of the future reformulated around the idea of thoughts, driven by the power of feelings, as shapers of our reality. When a cardiac patient visits a doctor, *the first prescription the doctor might offer might not be a drug, but a precisely formulated sequence of thoughts and feelings* designed to affect the genetic predisposition of people at risk for heart disease.

James Dillard, M.D., who told his patient Irv's story in *The Heart of Healing,* is a specialist in pain medicine, and director of the Pain Medicine Clinic at Beth Israel Medical Center in New York City. He wrote a compendious book on the subject that combines both alternative and allopathic approaches, called *The Chronic Pain Solution,*[2] and also wrote *Alternative Medicine for Dummies.*[3] He has been featured in *Newsweek* and *People* magazines, and has appeared on *Oprah*, NPR, and the *Today Show.*

Dillard's inspired impulse to employ his prescription pad to write down the name, not of a drug, but of a spiritual exercise, may have done far more for Irv than any pharmaceutical. It engaged the customized pharmacopeia of Irv's own immune system, a resource uniquely targeted to solve Irv's body's particular problems. Doctors are realizing that there are miraculous healing effects that occur as a result of changes in consciousness such as belief, intention, spiritual practice, and prayer. The medical profession is taking increasing note of how effective prayer is as a medical intervention; prayer is even becoming pervasive in the medical community.

A large-scale study was performed by the Jewish Theological Seminary in December of 2004. It surveyed 1,087 physicians. Among the doctors were practitioners of many faiths: Catholics, Protestants, Jews (broken out into groups of Orthodox, Conservative, Reform, and culturally identified but not religiously observant Jews), Muslims, Hindus, and Buddhists.

Doctors Often See Miracles

According to the results of the survey, two-thirds of doctors now believe that prayer is important in medicine. Miracles occur today, according to three-quarters of the group. Among physicians in every religious group in the study, except for less religiously observant Jews, more than 50% of participants believe that miracles occur today. Certain groups of doctors (among them Christians of all denominations, and Orthodox Jews) believed to a very high degree (80% or more) that miracles happen today.

Two thirds said that they encouraged their patients to pray, either because they believed it was psychologically beneficial to the patient, or because they believed that God might answer those prayers, or both. Half of them said that they encouraged their patients to have other people pray for them. Half of them said that they prayed for their patients as a whole, and almost 60% said that they pray for individual patients. An average of 55% of the physicians reported seeing miraculous recoveries in patients, and a third or more of physicians (of every religious group) said they had seen miraculous recoveries—even when the percentage of doctors in that group who prayed for patients was well below one-third.[4] Prayer has arrived in the consulting room and hospital in force—and perhaps it never left.

One of the striking findings of this poll is that between 50% and 80% of physicians—even those of weak religious faith—believe that miracles can happen today. There are many accounts of sudden and dramatic improvement in health, but it is not a well-studied phenomenon, since these tend to be viewed as anomalous phenomena. In their book, *Catastrophe Theory,* Alexander Woodcock and Monte Davis note that, "The mathematics underlying three hundred years of science, though powerful and successful, have encouraged a one-side view of change. These mathematical principles are ideally suited to analyze—because they were created to analyze—smooth, continuous quantitative change: the smoothly curving paths of planets around the sun, the continuously varying pressure of a gas as it is heated and cooled, the quantitative increase of a hormone level in the bloodstream. But there is another kind of change, too, change that is less suited to mathematical analysis: the abrupt bursting of a bubble, the discontinuous transition from ice at its melting point to water at its freezing point, the qualitative shift in our minds when we 'get' a pun or a play on words."[5]

Discontinuity and Transformation

A serious attempt to collect stories of sudden, discontinuous personal change has been made by psychiatry professor William Miller, Ph.D., at the University of New Mexico, author of some twenty-five books and many articles, and clinical psychologist Janet C'de Baca, Ph.D. After a newspaper article about their research in rapid personal shifts, they received hundreds of phone calls from people who had undergone rapid personal changes, including miraculous healings. They coined the term *quantum change* to describe this phenomenon, a term which, unlike the word "miracles," frees this type of experience from identification solely with religious observance (though the majority of quantum changes do indeed occur in the context of religious experiences). They describe many of these cases in their book, *Quantum Change.*[6]

They tell us that, "A decade later, we have a reasonably good description of the phenomenon and full confidence that sudden, profound, and enduring positive changes can and do occur in the lives of real people. Lives are transformed utterly and permanently, as utter darkness suddenly gives way to a joyful dawn that had not even been imagined. It happens." In the book they also look for the commonalities associated with all such quantum healings. Although they grapple to explain how and why it happens, they eventually offer five different perspectives that may explain why quantum change occurs. But there is no doubt that it occurs; our difficulties in measuring it and describing it are the result of the early stage of scientific inquiry in which we find ourselves, not an invalidation of the phenomena we are studying. In his book *Sacred Healing,* Norman Shealy put it this way: "Science can measure the earth only in plain facts in terms of electromagnetics, but quantum physicists have theories that are compatible with that subtle part of the unmeasurable higher dimensions. The mind appears to be capable of transcending time and space...."[7]

We change constantly, and the principles used in energy medicine can nudge that change in life-affirming directions. In *The Private Life of the Brain,* Susan Greenfield, Ph.D., reminds us that, "We are not fixed entities. Even within a day, within an hour, we are different. All the time, experiences leave their mark and in turn determine how we interpret new experiences. As the mind evolves, as we understand everything more deeply, we have increasing control over what happens to us: we are self-conscious. But this self-consciousness itself is not fixed. ...It will ebb and flow..."[8] Researcher Karl Maret, M.D., offers a strikingly similar description of the flexibility of genetic code: "the genome is plastic and resembles constantly rewritten software code rather than being fixed hardware that you inherit at birth."[9]

"It will take a miracle to rescue us," people exclaim when situations are dire. We usually look for miracles when we are in extreme peril. In a universe where the miraculous is available to us every day, where discontinuous positive change is always an option, and in which science has given us the understanding that genetic changes are occurring within a few seconds of changes in consciousness, it is high time that we began looking for miracles as a first resort, not as a last resort. We can choose to take conscious control of writing the mental and emotional software that can make us healthier and more vibrant, knowing that we can catalyze miracles in our bodies by doing so.

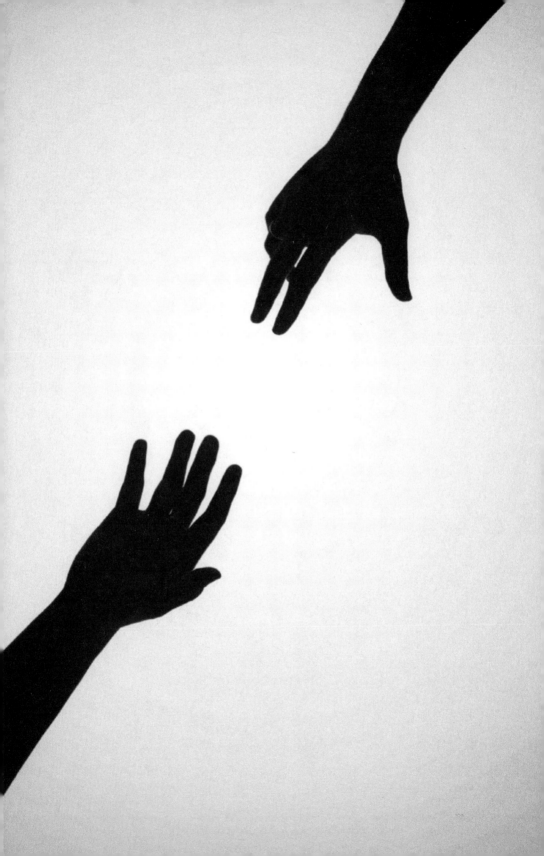

12

Energy Psychology

The fountain of content must spring up in the mind, and he who hath so little knowledge of human nature as to seek happiness by changing anything but his own disposition, will waste his life in fruitless efforts and multiply the grief he proposes to remove.

—Samuel Johnson, Elizabethan Lexicographer

"Michelle, a bright, perky, 21-year-old woman, arrived in my office complaining of severe bladder pain. She had to urinate frequently and urgently. I did a complete medical workup but could find nothing out of the ordinary—by the standards of my profession there was nothing wrong with Michelle. Yet it was clear to me that Michelle's pain was real, and her physical symptoms were real. After I had finished looking in her bladder with a cystoscope and found everything to be normal, I ventured, 'Sometimes women with your symptoms have a history of sexual abuse or molestation. Is this possible with you?' In the corner of her eye, the slightest of tears welled up. It turned out that Michelle had been sexually penetrated by an uncle almost daily from the age of three, till she was ten years old.

215

"I asked Michelle to think back upon these memories and find a part of her body where they were strongest. She said she could feel them acutely in her lower abdomen and pelvis. I asked her to rate them on a scale of 1 to 10, with 1 being the mildest and 10 being the most intense. Michelle rated her feelings at 10 out of a possible 10.

"I then spent 45 minutes working with Michelle, using some simple yet powerful emotional release techniques. I then asked her to rate her level of discomfort. It was a 1—complete peace. I urged her to cast around in her body for the remnants of any of the disturbed feelings she had previously felt. She could not find them, no matter how hard she tried. The emotionally charged memories had been so thoroughly released that a physical shift had occurred in her body. Her bladder condition disappeared. In the three years since that office visit, it has never once returned."[1]

Eric Robins, M.D., who tells this story in *The Heart of Healing*, is a urologist at Kaiser Permanente in San Diego, California. He is trained in several Energy Psychology techniques—powerful therapies which apply the principles of electromagnetic fields to medicine. All work with the body's electromagnetic signaling system to produce healing in the emotions and cells directly and quickly, without the need for extended courses of therapy, and sometimes without even needing to identify the experiences that caused the disturbance.

As Oschman explains, "stored trauma can be resolved as quickly as it was set in place. The body is continuously poised to resolve these afflictions and all of the physiological and emotional imbalances they create. This process goes to the deep energetic level that organizes or incarnates or underlies conscious experience itself. When this happens, the patient may suddenly know that the issue or discomfort will not bother them again."[2]

The first of these Energy Psychology techniques to be developed was Thought Field Therapy (TFT), and the most widely used is Emotional Freedom Techniques (EFT). Others include the Tapas Acupressure Technique (TAT), TEST, and WHEE (Wholistic Hybrid derived from EMDR and EFT). These therapies have demonstrated the ability to heal, in very short periods of time, psychological conditions that can take many months or years with conventional psychiatry and psychotherapy, if indeed they can be healed at all by conventional methods.

Emotional Freedom Techniques

Emotional Freedom Techniques (EFT), used by thousands of therapists, doctors, and lay people worldwide, has emerged as the most widely used of this class of therapies. EFT was developed by Stanford-trained engineer Gary Craig in the 1990s. After studying the existing methods, especially Thought Field Therapy, or TFT, he set out to find the simplest possible form for utilizing them. The result was a technique that did not require a therapist (though experienced therapists are useful in complex cases) and can be self-applied.

EFT consists of a simple routine. It starts with an affirmation: "Even though I have _____ (this problem), I fully and completely accept myself." While saying this, the participant taps a specific acupressure point with the tips of their fingers. Thirteen other acupressure meridian end points are then tapped five or more times by the participant with their fingertips, and a brief sequence of eye movements is undertaken. The whole procedure takes about forty seconds. Although it does not require a clinical setting, and can be self-administered, sophisticated uses for EFT do require training. The basics, however, are easily learned by anyone and can be applied to an impressive list of issues. In Gary Craig's words, "A dedicated twelve-year-old can achieve a 50% success ratio, even with problems that often stump doctors and conventional psychotherapists. Truly skilled practitioners often achieve 90%." EFT includes a self-assessment system that the subject can use before and then again, after the session, to determine whether or not the problem has been cleared with that session, or whether further work is required.

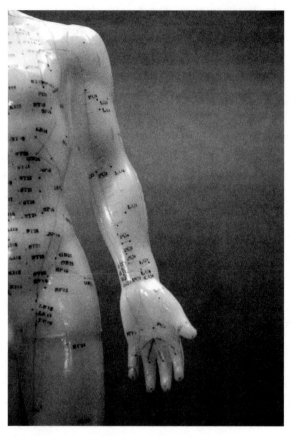

Energy meridians and acupuncture points

In a study of EFT published in the *Journal of Clinical Psychology* in 2003, Steven Wells, Ph.D., and his fellow researchers performed a controlled experiment to find out whether EFT could assist people with phobias. The subjects of the study had all been clinically diagnosed as having a phobia to small animals such as spiders, snakes, bats, and mice.

The study used several different measures of the strength of the participants' phobia before and after the study. They measured their pulses, to see the extent by which they rose when contemplating the object of their fear. They measured the number of steps toward the feared animal a participant could take. They gave participants written questionnaires and then gave the people in the experimental group a

30-minute explanation of the method, which included one brief session of EFT. The subjects were then put through the same tests.

The results were remarkable. On every measure, the subjects' fear dropped dramatically, and some were able to walk right up to the very animals that before had triggered crippling phobias. Not only were the results of EFT dramatic at the time, but in a follow up study done six months later, the subjects still had a much lower rate of phobic reaction to the objects of their fear.[3] Steve Wells recounts the following story:

> One of the ladies in the study was so afraid of mice that if she saw a mouse or rat, or even thought there was a mouse in the house, she would spend the night sleeping in her car! Needless to say, during pre-testing, she was unable to even enter the room with the mouse.
>
> After treatment, she was able to go right up to the mouse in the container without fear. Meanwhile, not long after the treatment, her daughter bought her granddaughter a pet rat as a present! A television station recording a program on our research went to this lady's home and took some footage of her. She was cradling the rat, and saying that they aren't such bad creatures after all!
>
> There were many similar examples of people who also couldn't enter the room who, after just 30 minutes of EFT treatment, were able to not only open the door but go right up to the creature that would previously have caused them to run from the room. One of these, a lady with a cockroach phobia, felt that her phobia had been holding her back in many areas of her life. Following 30 minutes of an EFT treatment after which she went straight into the room and picked up the cockroach in the jar to examine it closely, she reported a huge shift in self-esteem and confidence that permeated all parts of her life.[4]

Steven Wells's study was later replicated and extended by another research team led by Harvey Baker of New York's Queen's

College, lending further credibility to the results.[5] And the pilot for a new movie being filmed by the producers of the hit movie *The Secret* captures footage of similar scenes. In one shot, a woman with a life-time fear of cockroaches panics and screams uncontrollably when her attention is drawn to a fake cockroach. After a brief EFT treatment, she is able to calmly handle a jar containing a live cockroach.

For any therapy to be able to produce such large results in so small a time is almost mind-boggling. When I first heard of the results being obtained through EFT, I dismissed them as hyperbole; I simply could not believe it. However, after having tried the technique myself, on some mental conditions where other approaches like meditation were having no impact, I was impressed with the immediate changes I experienced. After learning more about the technique from Gary Craig, the originator of EFT, and witnessing the spectacular and immediate results reported by his subjects, I began to apply it with other people. The case history with which I begin this book is an example of EFT, as are many of the case histories that I quote in other chapters. While the routine is very simple, applying it effectively to sufferers from long-term psychological trauma requires a deep understanding of its applications and limitations.

Therapists have noted that people under stress often touch, tap or rub certain parts of their bodies. Examples include:

Rubbing the temples on either side of the eyes
Putting a hand to the mouth, with fingers touching the skin above the upper lip and below the lower lip
Raising a hand to the heart area
Wringing the hands
Fidgeting with the fingernails with the opposite hand
Resting the head on a hand supporting the bridge of the nose

It turns out that specific areas of the body, when tapped or rubbed, release stress. But when they are *all tapped or rubbed in turn, the stress reduction effect is amplified.* Energy Psychology takes these acu-points out of the realm of *unconscious self-interventions* and *systematizes the conscious use* of these points into an organized stress-reduction routine. Once memorized, the routine can be used for reducing or eliminating stress whenever it occurs, and even for releasing stress that has long been buried in your muscles or your subconscious mind.

Unconscious stress-reduction self-interventions

We've all had the experience of buying a new car and suddenly noticing how many of them there are on the road. A few years back, I purchased a gold Dodge Caravan so that my children and I could go camping. My son Lionel, then 15 years old, was awed by the car. "Rare" was one of his favorite words. "This car's so cool," he exclaimed. "It's rare!"

Later he saw another one driving down the road, and was amazed. "Look, Dad, there's another car just like ours!" We drove into a parking lot. There were two other gold Dodge Caravans in the lot. We parked next to one of them. Lionel was crestfallen.

Observation of stress-reduction points is like that too. Once you become aware of them, you often notice yourself and others touching them unconsciously. Take a look at the accompanying photos. You've probably often seen people reacting to stress, shock, or emotional discomfort by touching one of these points. They are so ubiquitous that they're usually overlooked, and up until now psychologists and biologists have had little idea that these common gestures are part of our instinctive response to stress.

Stress Relief and EFT

In a study published in *Counseling and Clinical Psychology Journal,* psychologist Jack Rowe, Ph.D., of Texas A&M University's Psychology and Sociology department, tested people for stress before and after EFT. He administered standardized test for measuring stress, called the SLC-90-R, to 102 people before, during, and after an EFT workshop with Gary Craig. Rowe measured their stress levels one month before the workshop, when the workshop started, when it ended, one month after the workshop, and six months later. He found that the stress levels of participants decreased significantly between the beginning and end of the workshop. But even more promising was the finding that the effect endured, and that when subjects were re-tested six months later, their stress levels remained much lower than they had been before the event. Such was the strength of the result, that there was less than one chance in two thousand that such results could have occurred through chance.[6]

Another pilot study of EFT examined its effects on patients about to undergo dental surgery. Research confirms that about one in three people experience moderate to severe anxiety when confronted with dental treatment. After EFT, 100% of patients in the study reported a decrease in anxiety.[7] Another study examined the EEG patterns of claustrophobic patients when they were confined in a small, metal-lined enclosure similar to an elevator, and compared them with a control group. It found that they had high concentrations of Theta waves, as well as getting high scores on an anxiety test. After a TFT treatment, their brain wave profile returned to the same level of Theta activity as the non-claustrophobic subjects, and their scores on the anxiety test also dropped. Their anxiety scores also dropped. Two weeks later, the results of the experiment still held steady.[8] Another studied the effects of acupressure on subjects who had received a minor injury that required paramedics to transport them to hospital. Those who had received acupressure showed a significantly greater reduction of anxiety, pain, and heart rate than those who had not.[9] While these pilot studies are not definitive, they suggestive fruitful directions for further research.

Researchers Study Energy Psychology

Initial acceptance of Energy Psychology toward the end of the twentieth century was hindered by the lack of a systematic theoretical framework to explain their efficacy, as well as the lack of an experimental database of studies demonstrating their effectiveness. But in the last ten years, several experiments using Energy Psychology have been reported in scientific journals, and the efficacy of these treatments means that more will follow.[10] The scientific evidence for the field is being tracked by several institutions like the Energy Medicine Institute, and they update their sources as new papers are published.[11]

Researchers are now understanding not just *that* they work, but *how* they work. The probability is that *tapping creates a piezoelectric charge that travels through the connective tissue along the path of least electrical resistance. When coupled with conscious memory of a trauma and an awareness of the site in the body that holds the primary memory of the trauma, the IEGs that are implicit in healing are activated, and the intensity of physical feeling at the site is discharged, taking with it the intensity of emotion related to the trauma.*

In one experiment, the brain scans of subjects suffering from generalized anxiety disorder were examined. Conditions such as anxiety and depression "can be distinguished by specific brain-frequency patterns. Anxiety has one such electronic 'signature'.... Depression has another,"[12] according to David Feinstein, Ph.D., coauthor of *The Promise of Energy Psychology,* which describes this study in fascinating detail.

Digitized EEG readings of subjects' brain scans were taken before treatment began. They then received twelve energy treatment sessions, and a second EEG was taken. The group receiving Energy Psychology treatment was compared to a group receiving the Gold Standard in experimentally validated "talk" therapy, cognitive behavior therapy (CBT). It also compared them with patients receiving medication. The patients were interviewed three, six, and twelve months after treatment to determine whether the therapy had produced lasting results.

The EEG readouts demonstrate the regions of their brains that are functioning normally, and those with various degrees of dysfunction. The illustration below indicates the progress made by those subjects that received EFT. It shows that before treatment began, most areas of their brain revealed high or very high levels of dysfunctionality. Only a small portion of their frontal lobes shows normal or slightly dysfunctional patterning.

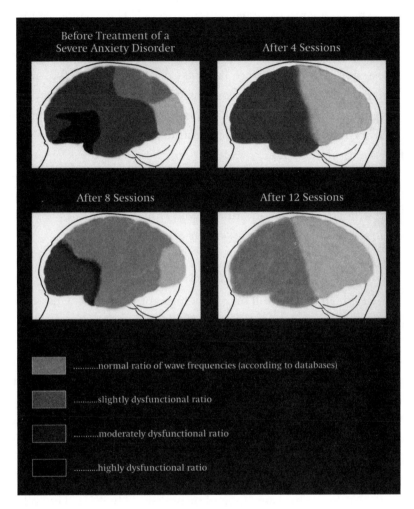

Brain scans over the course of twelve energy treatment sessions[13]

After twelve sessions of energy treatments, as can be seen on the right bottom scan, most areas are normal or close to normal. Patients given CBT had similar results, but "more sessions were required to accomplish the changes, and the results were not as durable on a one-year follow-up as were the energy psychology treatments." As Oschman notes, citing the findings of William M. Redpath, "no matter how well patients recalled critical memories, the patterns of their lives continued to be tormented and disrupted. The most difficult issues...must be resolved by the organism itself."[14]

And compared to giving patients drugs, "The brain-wave ratios did not change, suggesting that the medication suppressed the symptoms without addressing the underlying wave-frequency imbalances. Undesirable side effects were reported. Symptoms tended to return after the medication was discontinued."[15] When symptoms are suppressed, the patient may act and feel more normal. But the underlying brain rhythms remain unaffected. Antidepressant drugs are *masking* their symptoms, without *removing* them. This is like taking a painkiller when you have a broken arm. You might *feel* better, but left to fester, your *underlying* condition remains unchanged. As Shakespeare said, "It doeth but skin and film the ulcerous place, whiles rank corruption, mining all within, infects unseen."

Energy Cures for Physical Traumas

Besides working well on psychological traumas, EFT has established a surprising track record of effectiveness in shifting, or even curing physical ailments. This is because there is virtually always an emotional component of a physical disease.[16] Cancer patients, for instance are often psychologically depressed, along with the depression of their immune system that accompanies chemotherapy. Psychological depression left untreated often contributes to organic disease. Even if a patient has a disease that seems purely organic, such as an acute bacterial infection, releasing any possible emotional stressors cannot hurt. I have been surprised to find that even colds and flu are sometimes ameliorated or prevented by EFT. The following story illustrates the healing of a physical ailment after using EFT:

> A woman who had suffered from carpal tunnel syndrome for about eight years received some training in EFT. After having had extensive chiropractic and physical-therapy treatment, she thought she was about as healed as she was going to be without surgery. Even though she felt better in general, the chronic pain still wore her down occasionally. After being introduced to EFT and being guided through a tapping sequence for about ten minutes, she had absolutely no pain! She reports, 'I couldn't believe it! I kept mentally searching throughout my body for

the pain. It just wasn't there! I had actually taken a few minutes to tap on the corner of my eye, my hand, and other simple, easy-to-do places on my body and had shed this pain with which I thought I was destined to live!' Next she applied it to grief that had burdened her for many years. Within minutes, she recounts, 'I felt the weight lift from me. I have been able to remember my sister fondly and nostalgically, but with no pain of grief, ever since!'[17]

I have seen the same results from EFT time and time again. While writing this chapter, a friend of mine who graciously tends my garden came in sniffling. "I have allergies," she explained.

"What's the worst?" I inquired.

"Grass seeds," she said without hesitation. A couple of minutes after I asked her to intensify the feeling, she frowned, and said she had a headache. The allergy and headache was a 7 on a scale of 10, she said. Her nose was so clogged that she sounded stifled.

I did an EFT tapping routine on her. She said she was a zero.

I ran out the door, cut a few stalks of flowering grass, came back in, and presented her with the bouquet as a humorous tribute to healing. While she held them, I did another tapping routine. Then I asked her how she felt, on a scale of 0 to 10.

"I'm a zero!" she said. Her eyes opened wide at the realization. But the sound of her voice showed that more than her eyes had opened: her nasal passages didn't sound at all clogged as she spoke. She sat on the chair for a while longer, dazed, clutching the paradoxical bouquet in amazement. In the words of a leading researcher, what is at work in these cases is that they allow for, "organized or non-chaotic energy to spread suddenly throughout the organism to create new structures, functions, and order. This concept is important as a frequent observation of practitioners of energy psychology, bodywork, energetic and movement therapies is a sudden and beneficial 'sea change' or 'phase change' spreading throughout the organism as trauma or other disorder is resolved, and the whole body reintegrates accordingly."[18]

One doctor examined her patient's live red blood cells using dark-field microscopy before and after the patient used EFT. The doctor was worried about the degree of clumping found in the patient's blood sample, since when red blood cells clump together, they present less surface area to absorb oxygen from the lungs and distribute it throughout the body. An even distribution of red blood cells indicates a healthy ability for oxygen absorption, while clumping indicates decreased oxygen distribution.

She found that the patient could alter the degree of clumping of her red blood corpuscles by doing EFT, and that the effect showed up immediately when the samples of live cells were examined under a microscope. The first photo shows the patient's red blood cells clumped up before doing EFT to release them. The second photo, taken a few minutes later, shows the patient's red blood cells after two rounds of EFT done with the conscious intent of producing even distribution of cells.[19]

Our dominant medical model does not believe that physiological processes such as red blood cell clumping are under the control of a patient's conscious mind. Reversing red blood cell clumping usually takes months of treatment, if it can be accomplished at all. Yet in this case, two rounds of EFT, lasting just a few minutes, were enough to produce an immediate and visible change in the degree of cell clumping. Stories such as this provide promising pointers for future controlled experiments in the effects people can produce in their bodies by using EFT.

The EFT website contains stories by hundreds of doctors, psychiatrists, psychotherapists, sports coaches, social workers, and other health professionals of ways in which they have found it useful, and of patients who have responded to EFT after conventional interventions had failed. And the EMDR websites list some thirty clinical studies demonstrating the effectiveness of this therapy in clinical trials.[20]

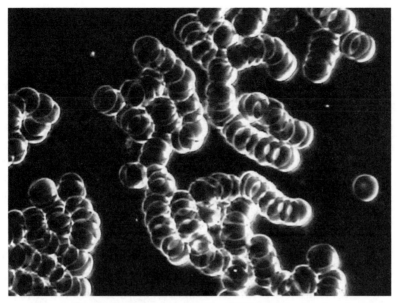

Red blood cells clumped before EFT for cell release

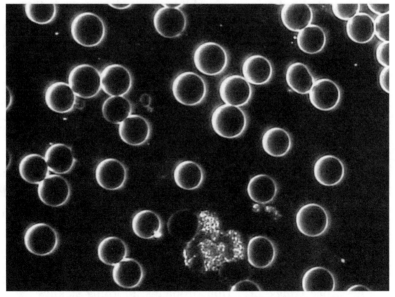

Red blood cells evenly distributed after EFT for cell spacing

EFT and other Energy Psychology techiniques give us access to the realm of quantum healing, and promise to dramatically shorten the time taken to heal psychological traumas. To professionals

229

accustomed to lengthy courses of psychotherapy, or to resorting to drug therapies, they indeed seem like miracles.

An excellent guide to the EFT routine, called the "Basic Recipe" by authors David Feinstein, Donna Eden, and Gary Craig, is found in their book *The Promise of Energy Psychology*,[21] and also appears as an appendix to this volume. In their account, they list some of the conditions that have been shifted or cured by EFT:

> The following are actual examples that illustrate the range of issues where simply applying the Basic Recipe gave someone relief: performance fears for a nineteen-year-old gymnast, flashbacks and insomnia a woman was experiencing following two automobile accidents during a six-week period, a refinery worker stopping smoking after thirty-five years, a woman's extreme anxiety prior to bladder surgery, a six-year-old girl's psychosomatic pains, a mother's fear of flying that was being communicated to her one-year-old daughter, depression suffered by a single mom with two teenage daughters, a woman's intense lifelong craving for chocolate and ice cream, a thirteen-year-old boy's fear of the dark, a boy with an intense allergic reaction to horses, another boy with severe dyslexia, a woman's pain after reconstructive surgery for a damaged knee. You can read details about each of these examples, as well as hundreds of others, at www.emofree.com.[22]

Reports found on the www.emofree.com site suggest that basic tapping methods have resulted in improvement in headaches, back pain, stiff neck and shoulders, joint pains, cancer, chronic fatigue syndrome, lupus, ulcerative colitis, psoriasis, asthma, allergies, itching eyes, body sores, rashes, insomnia, constipation, irritable bowel syndrome, eyesight, muscle tightness, bee stings, urination problems, morning sickness, PMS, sexual dysfunction, sweating, poor coordination, carpal tunnel syndrome, arthritis, numbness in the fingers, stomachaches, toothaches, trembling, and multiple sclerosis among many other physical conditions. This is not to suggest that the Basic Recipe replaces medical care, but it is interesting that an approach designed to address emotional problems is so frequently reported as helping with physical problems as well.[23]

Ever-Quicker Interventions

Psychiatrist Daniel Benor, M.D., is one of the pioneers of mind-body therapy. In the early 1980s, he was the first person to compile a comprehensive database of research studies supporting noetic interventions. He has developed a hybrid form of meridian therapy, drawing on the same principles that underlie EFT and EMDR. He calls his technique WHEE (Wholistic Hybrid derived from EMDR and EFT). WHEE takes less time than EFT, in case your life is so fast-paced that the ninety seconds required for an EFT routine seems like an eternity to you!

You begin a WHEE intervention by thinking about the issue, or while you're in the middle of a stressful situation. You then assess yourself on a scale of zero to ten, with zero being completely calm, and ten being "fit to be tied." You can also focus on the location in your body where the feeling is most intense.

You begin with an affirmation similar to one you might use for EFT, such as, "Even though I feel _____ (e.g. "dreadful when I think of Barbara hitting me"), I love myself wholly and completely, and God loves me wholly and completely and unconditionally." You then alternate tapping the two sides of your body in some manner. You can tap the insides of your eyebrows where they join the bridge of your nose, or the edges of your eyes. If you feel resistance to doing the exercise, rub just below the collarbone center point to release it.

You can also squeeze your toes. First you squeeze your left big toe down against the ground, then your right big toe down, and so on.

You then anchor a positive affirmation, while continuing the alternate tapping. Think about a benefit from the situation, like:

I'm safe now.

I've connected with some wonderful people.

I've learned to ask for help.

I've learned boundaries.

You then assess yourself again on a scale from one to ten, and repeat the alternate tapping procedure if you don't feel considerably better. The routine takes around thirty seconds.[24]

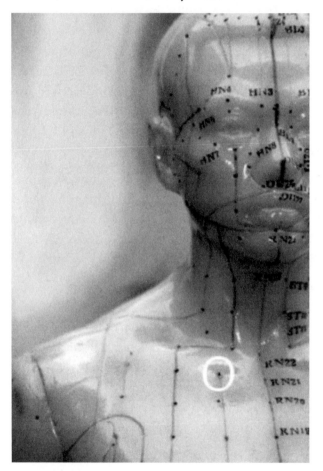

Collarbone center (indicated by white circle)

Covert Relief

Beyond its speed, however, WHEE has the benefit of not *looking* obvious. When you're in the courtroom in the middle of a divorce trial, when you're in a meeting with the boss, when you're on a hot date, when you're gearing up for a television appearance, when you're in the middle of a fight with your spouse, or when you're on the sports field playing in front of thousands of spectators, you can't call a time out and start tapping your temples and rolling your eyes without danger of arrest. That's where WHEE comes in: you certainly *can* tap your toes without looking like anything other than the impatient person you probably are!

A population with whom Daniel Benor has had great success is young people who are locked up in juvenile hall. He says that teenagers usually refuse to use tapping outside of his office, for fear of embarrassment. They can't risk looking ridiculous in front of their peers. He has therefore developed ways in which they can reduce their tempers and their triggering with WHEE without anyone around them noticing. Besides tapping the toes, he has experimented with tapping alternate sides of your teeth with your tongue, and other subtle alternating points. The potential for defusing dangerous conflicts by these fast and easily learned techniques is substantial. The following case history is a composite history drawn from Dr. Benor's web site:

> Six-year old "Joe" had been seriously abused emotionally, physically and sexually by his mother from at least the age of two and probably earlier. He was removed from her home at age four, and had nine foster home placements before his latest foster mother suggested to the welfare worker that counseling might be helpful to him for his temper outbursts, fighting children in his first grade class and after school, inability to fall asleep till past midnight, frequent nightmares and night terrors, and bedwetting—his more serious problems. In addition, he was unable to sit still, was impulsive, distractible, constantly forgetting and losing things, and had no friends.
>
> I diagnosed PTSD (moderately severe) and possible attention deficit hyperactivity disorder (ADHD). I prescribed small doses of Ritalin, as this acts within minutes and, if effective, could provide rapid relief for some of his problems.

He responded well to the medication and was much better able to sit and attend in class, less impulsive, and less forgetful. His other symptoms remained. He also had counseling sessions weekly with a social worker at the clinic where I work, focused on issues of relating to his new family, multiple losses, and PTSD issues.

At the initial interview, I taught Joe and his [foster] mother to use the butterfly hug [crossing both arms over the chest, hands draped over the front of the opposing bicep]. Joe chose an affirmation about one of the bad memories he had of being left in the dark cellar by his [birth] mother. He was unable to count, so I had him show me a Visual Analog Scale ('VAS'—equivalent of a SUDS) with the gap between his hands representing how big his bad feeling was when he thought about being in the dark cellar. He opened his arms wide and said, 'I can't reach to show you how big the bad feeling is.' Within minutes of using the butterfly hug, his hands were touching in the VAS. He had reduced the bad feelings to zero.

Over the next several weeks, Joe (with the help of his mother) used the butterfly hug daily on various fears, difficulty falling asleep, and nightmares, as well as to calm down after he had temper outbursts.

Within two months, Joe was functioning at near-normal levels of behavior in school and at home. Counseling continued for another four months and was discontinued. I have followed him for Ritalin prescriptions for two years and we have all been pleased with his excellent academic progress in school, and with his good behavioral and improved social adjustments in school and at home.[25]

Tapping works with babies as well as older children, according to experienced clinicians, Roberta Temes, Ph.D., reports, "I've seen babies stop crying in mid-sob when the tapping begins. In an informal experiment in a day care center in New Jersey, I observed one group of nannies lift crying babies to their laps and speak soothingly to them, while other nannies lifted their babies, soothed them with words, and also tapped.... The latter group all stopped crying. The former group, offered comfort but no taps, did include several babies who slowed down their crying, but some continued crying and only one totally stopped.[26]

There are many accounts on Dr. Benor's website of adults as well as children who have released and resolved long-standing traumas, often after conventional psychotherapy has failed to produce relief. This is Third Stage medicine: prayer, intention, meditation, visualization, belief, Energy Psychology and other "soft" disciplines that could someday make much "hard" medicine, such as invasive surgery, largely obsolete. They provide, says Lipton, "the scientific underpinning for pharmaceutical-free energy medicine."[27] Apparently we exist in a sea of quantum potentials, and we affect which ones are actualized by the quality of our feelings and thoughts. The HeartMath researchers express the hope that "understanding gained through this process could...form the basis for a new mode of treating diseases and disorders that are largely unresponsive to existing medical treatments."

Changing the Beliefs Below the Symptoms

I recently attended a seminar that taught a method for healing subconscious beliefs called PSYCH-K. This technique, developed by therapist Rob Williams, determines which beliefs are held in your subconscious mind, and then shifts them. Participants troll through lists of beliefs that are embedded in their belief structure such as:

I give myself permission to do what I love.
It's okay for others to disagree with me.
I am able to function independently of others.
It's easy for me to receive love from others.
I experience the presence of God within me.
Money is my friend.
My body heals itself naturally and quickly.
It is safe, fun, and easy for me to be slim and healthy.

Rather than focusing on conscious beliefs—everyone believes in wealth, for instance, even poor people—Williams tests the strength of muscles when the subject makes a statement. If a muscle tests weak when making a statement, Williams believes that our subconscious minds don't believe it, even though our conscious minds might consciously affirm it. If our muscles test strong, it indicates congruence between the conscious statement and the unconscious belief.

The beliefs for which Rob Williams tests are in six areas challenging to most people: health and the body, prosperity, relationships, self-esteem, personal power, and spirituality. As they run through the lists, participants notice the statements for which their muscles test weak, and then work with them using PSYCH-K's somatic healing tools. Once a negative belief has been healed, the person will test strong when confronted with its positive counterpart.

For instance, I tested strong for most of the twenty-five beliefs about money and prosperity. But I tested weak for two:

I can afford to take time off to rest and nurture myself whenever I need to.
Money is one expression of my spirituality and my love for God, myself, and others.

I knew before the test that those two statements would render me weak! My whole body felt weaker simply reading the statements. I hadn't considered that money could actually be an expression of my love for God. And I often get impatient with my needs for self-nurturance when I see so much suffering in the world, and I can play a small part in alleviating it. The list of beliefs allowed me to pinpoint exactly where my energy was not flowing. It showed me the beliefs I held that were subtly anchoring me to unhealthy habits and physical conditions.

For one of the exercises on the first day of the seminar, I partnered with a woman who came in on crutches. She had suffered for fifteen years, she told me, from a debilitating neurological condition that arose after she was injured in a car accident. Her broken bones had healed, but the pain was so intense that she took large doses of painkillers three times a day, and one at night to allow her to sleep. She had been on this high dosage of painkillers for every day of the past fifteen years.

On the final day, as people were walking around saying goodbye and giving each other hugs, I noticed she was not hobbling around on her crutches. She told me that she felt fine, and that she had in fact felt so good the previous evening that she had forgotten to take her nightly dose of medication. She woke up that morning feeling even better, and had not taken painkillers the entire day. It was now twenty-four hours since her last dose. And she was walked around the whole day without once resorting to her crutches.

There are hundreds of such stories in the world of PSYCH-K. A former president of the Colorado Association of Psychotherapists wrote: "I was scheduled to speak at a local high school and, as an afterthought, was invited to spend 45 minutes with the track team. I gave them a twenty-minute lecture on 'beliefs and limitation' and then did a demonstration of PSYCH-K using one of their sprinters who was having trouble getting out of the blocks. He was noticeably improved in just minutes and that got the rest of the team's attention. All I had time to teach them was how to muscle test their beliefs about winning and how to do the Whole Brain Posture. At the next meet, two days later, they broke 15 out of 22 school records, one state record, and a sixteen-year-old national girl's high school relay record by five seconds! PSYCH-K is clearly the most amazing tool for effective personal change that I have ever encountered."[28] Other testimonials report cessation from various fears and phobias, allergies, depression, weight problems, and a variety of organic diseases.[29]

As Easy as TAT

An easily learned technique that can be applied to a wide variety of anxieties and traumas is TAT, or the Tapas Acupressure Technique, named after its founder, Tapas Fleming.

In TAT, you place three fingers of either hand on the front of your face, and the flat of your other hand at the base of your skull. You touch the ring finger to the thumb, then place the tips of both fingers on the inside edges of the eyes, on either side of the bridge of the nose. There's no need to pinch or apply pressure. The long finger is then placed gently on the forehead. The other hand goes flat on the back of your head, on the bony protuberance where the skull terminates.

Tapas Acupressure Technique (TAT)

You hold this pose while you think of the worst part of the problem that's bothering you, and wait till you feel a physical shift. It usually happens in a minute or two, rarely longer than five. You feel your body relax as the tension drains from you.

Then, while holding the position, you anchor a positive affirmation, something like, "All the origins of this problem are now healed," or, "Everything that led up to this situation happened, but now it's over, and it's no longer resulting in pain," or, "All the sources of this problem in my mind, body, heart, life, and all other dimensions of my being are healed now."

Again, you will feel a distinct physical shift as the affirmation sinks in.

At a tutorial I did with Tapas Fleming, one of the questions an audience member asked her was instructive. "Isn't it healthy to *process* emotion?" she asked; "Isn't there value to getting to the roots of the problem through processing, and stay with that process till it's resolved?" This question might be asked of all Energy Psychologies, because they abort the usual long process of psychological enquiry so familiar to psychotherapists, and go straight to healing.

Fleming's point of view was intriguing. "You don't have to process emotion," she responded. "Emotion comes from a particular point of view, a particular identity. When you're invested in that point of view, that identity, you're going to feel the associated emotion.

"But emotion pours out of that identity continually. If you take up that point of view, all the emotions that accompany it will pour out of you as long as you stay there. If you shift your point of view, you no longer have to process the emotions the flow from that identity."[30]

While these Energy Psychology techniques are relatively new, and seem exotic to the eye accustomed to the clinical paraphernalia of Western medicine, they can also be seen simply as modern applications of the ancient energy pathways familiar to the Oriental *qi* culture of two thousand years ago. Tapas Fleming recounts hearing of an ancient acupuncture text that contained a dialog between a student and a master. The student asks, "I hear that in ancient times, doctors didn't have to use acupuncture and herbs."

The master replies, "In the old days, energy was transmitted directly." Fleming speculates, "Acupuncture and herbs may be a step down from direct transmission of energy. And Energy Psychology may merely be a rediscovery of the ancient methods of direct transmission."[31] Larry Stoler, Ph.D., president of the Association for Comprehensive Energy Psychology (ACEP), succinctly observes, "Energy Psychology is applied qigong."[32]

The direct transmission of healing also has its place in the Christian tradition. During a busy day of ministry, thronged with crowds, Jesus was aware of direct transmission to a single individual:

> And a woman was there who had been subject to bleeding for twelve years, but no one could heal her. She came up behind him and touched the edge of his cloak, and immediately her bleeding stopped. "Who touched me?" Jesus asked. When they all denied it, Peter said, "Master, the people are crowding and pressing against you." But Jesus said, "Someone touched me; I know that power has gone out from me." Then the woman, seeing that she could not go unnoticed, came trembling and fell at his feet. In the presence of all the people, she told why she had touched him and how she had been instantly healed. Then he said to her, "Daughter, your faith has healed you. Go in peace."

King Charles II laying on hands

The woman who touched the hem of his garment evidently believed in direct transmission of healing grace, and apparently the Master did too, because he felt it flow from him. England's King Edward the Confessor touched many subjects who reported healing, and Charles II is reported to have bestowed the king's touch on around 4,000 people each year.

It is likely that prayer, intention, and belief strengthen the effect of these Energy Psychology techniques. A researcher with Johns Hopkins University observed that, "intention, expectation, culture, and meaning are central to placebo-effect phenomena and are substantive determinants of health."[33] Any cultural or religious belief that boosts intention also boosts the beneficial effects of those factors on health. For instance, a friend of mine who is a devout Christian uses the Tapas technique in conjunction with traditional Christian affirmations of the healing power of the Trinity.

The three points touched on the forehead are in the same area as that traditionally touched by the priest or deacon while saying the words, "In the name of the Father, Son, and Holy Spirit."

Priest offering absolution of sins

My friend modifies the affirmation to say, "Even though I have this _____ (problem), I now receive divine grace, and complete healing, in the name of the Father, Son, and Holy Ghost." Using her faith to reinforce an energy medicine technique gives her the best of both worlds.

Near the start of the *Yellow Emperor's Classic of Internal Medicine,* the ancient Chinese text that identifies the energy pathways used by these modern techniques, the Yellow Emperor says, "I have heard that in early ancient times, there were the Enlightened People who could...breathe in the essence of *qi*, meditate, and their spirit and body would become whole."[34] When you start to put all the pieces together; the scientific studies of Energy Psychology, the thousands of case histories reported on the web sites of practitioners, and the research on the effect of belief and faith on healing, it is apparent that a huge arsenal of treatments is emerging that is safe, swift, and effective. We are no longer limited to a repertoire of drugs and surgery for our wellbeing. The new medicine also offers each of us a degree of control over our wellness, down to the very level of our cells—one that science never even dreamed of a generation ago.

13

Soul Medicine as Conventional Medicine

We have to recognize that we are spiritual beings with souls existing in a spiritual world as well as material beings with bodies and brains existing in a material world.

—Nobel Laureate Sir John Eccles[1]

"Tim Garton, a world champion swimmer, was diagnosed in 1989 with stage two non-Hodgkin's lymphoma. He was forty-nine years old and had a tumor the size of football in his abdomen. It was treated with surgery, followed by four chemotherapy treatments over twelve weeks, with subsequent abdominal radiation for eight weeks. Despite initial concern that the cancer appeared to be terminal, the treatment was successful, and by 1990, Tim was told that he was in remission. He was also told that he would never again compete at a national or international level. However, in 1992, Tim Garton returned to competitive swimming, and won the one hundred meter freestyle world championship.

245

"In early July of 1999, he was diagnosed with prostate cancer. A prostatectomy in late July revealed that the cancer had expanded beyond the borders of his prostate and could not all be surgically removed. Once again, he received weekly radiation treatment in the area of his abdomen. After eight weeks of treatment, the cancer had cleared.

"In 2001, the lymphoma returned, this time in his neck. It was removed surgically. Tim again received radiation, though this time it left severe burns on his neck. The following year, a growth on the other side of his neck, moving over his trachea, was diagnosed as a fast-growing lymphoma that required emergency surgery.

"He was told that the lymphoma was widespread. An autologous bone marrow and stem cell transplant was done at this time, but it was not successful. There was also concern that the tumors would metastasize to his stomach. His doctors determined at this point that they could do nothing more for him. He was told that highly experimental medical treatments, for which there was little optimism, were the only alternative. He was given an injection of monoclonal antibodies (Retuxan), which had been minimally approved for recurrent low-grade lymphoma. Retuxan is designed to flag the cancer sites and potentially help stimulate the immune system to know where to focus.

"At this point Tim enlisted the services of Kim Wedman, an energy medicine practitioner trained by Donna Eden. Tim and his wife went to the Bahamas for three weeks, and they brought Kim with them for the first week. Kim provided daily sessions lasting an hour and a half. These sessions included a basic energy balancing routine, 'meridian tracing,' a 'chakra clearing,' work with the 'electrical,' 'neurolymphatic,' and 'neurovascular' points, a correction for energies that are designed to cross over from one side of the body to the other but are not, and a daily assessment of his other basic energy systems, followed by corrections for any that were out of balance. While Tim was not willing to follow Kim's advice to curb his substantial alcohol consumption, or modify his meat-and-potatoes diet, he did introduce fresh vegetable juices into his regimen, plus an herbal tea (Flor-Essence) that is believed to have medicinal properties.

"Kim also taught Tim and his wife a twenty-minute, twice-daily energy medicine protocol, which they followed diligently, both during the week she was there, and for the subsequent two weeks. The protocol included a basic energy balancing routine, and specific interventions for the energy pathways that govern the immune system and that feed energy to the stomach, kidneys, and bladder.

"Upon returning to his home in Denver, in order to determine how quickly the cancers might be spreading, Tim scheduled a follow-up assessment with the oncologist who had told him, 'There is nothing more that we can do for you.' To everyone's thrill and surprise, Tim was cancer-free. He has remained so during the four years between that assessment and the time of this writing. He has been checked with a PET scan each year, with no cancer detected. Was it the energy treatments or the single Retuxan shot that caused the cancer to go into remission over those three weeks? No one knows. Tim still receives Retuxan injections every two months, but he also continues to work with Kim Wedman on occasion for tune-ups."[1]

While the names and other identifying characteristics of patients whose cases are described in this book have been concealed, in the above case they have not; Tim Garton gave his permission for this story to be published. Kim Wedman, the energy practitioner who worked with Tim Garton, trained with Donna Eden, author of *Energy Medicine,* and her husband, clinical psychologist David Feinstein, Ph.D.

Energy Medicine in Hospitals

The University of Texas has a cancer research and treatment facility in Houston called the M.D. Anderson Cancer Center. In collaboration with a seven-hospital chain called Orlando Regional Healthcare, it has set up the M.D. Anderson Cancer Center Orlando (MDACCO) in Florida.

M.D. Anderson Cancer Center Orlando

Besides offering a wide range of cancer therapies and excellent conventional care, M.D. Anderson set up the pioneering Mind-Body-Spirit department that includes an Energy Medicine Program. It was set up and is managed by Patricia Butler, M.A., and uses the techniques of Donna Eden's Energy Medicine program. Butler provides counseling and Energy Psychology services to the staff, and also provides energy medicine services to patients on request to reduce the side effects of chemotherapy and radiation. Information on the Energy Medicine Program is found in the Patient Information Guide given to every new cancer patient.

During chemotherapy treatments, Butler may sit with patients and do this work while they're hooked up to the chemotherapy drip. She has observed many beneficial effects in patients who combine energy medicine with their conventional treatments. Patients often report feeling much better during and after chemotherapy sessions where they simultaneously receive energy medicine treatments than chemotherapy sessions where they don't. "Energy medicine seems to calm the body's energy systems, enabling them to tolerate the presence of chemotherapy drugs," she observes. And they often don't require further medications for insomnia or listlessness. She wrote the following account of a typical case:

248

Sylvia was fifty-five years old when doctors discovered a malignant gastric tumor growing right at the joint between her esophagus and her stomach. With the shock of this news resonating wildly throughout her system, she listened as they recommended chemotherapy and radiation to shrink the tumor followed by surgery to remove what was left behind. Feeling nervous about the outcome, but not knowing what else to do, she followed all of their protocols and underwent the surgery as soon as it was indicated. The good news, she learned afterwards, was that the operation had been a success and she was now cancer-free; the bad news was that they'd had to remove her entire stomach to get all of the cancer. They placed a feeding tube in her side to ensure that she would get nutrition, and told her they hoped she could eat normal food one day (albeit with careful new strategies). Although she did her best to recover from both her treatments and the rearrangement of her digestive tract, her quality of life took a real nosedive. The next six months were simply miserable.

Sylvia's follow-up care was transferred to M.D. Anderson Cancer Center Orlando (not her original treatment center), and she learned about the hospital's Energy Medicine Program. She came to see me, wondering if energy medicine could help her begin to feel better and get her life back again. As she began to tell me about her very uncomfortable state, I learned that the worst part for her was the nausea that she'd felt every day since her surgery six months ago. Not only was it debilitating, it kept her from being able to eat, and she couldn't have her feeding tube removed until she could eat again. Along with the nausea, Sylvia still felt quite a bit of pain along the incision, at the tube site, and in her lower back. It took 20 mg of methadone three times a day to hold the pain at bay, and she didn't like the way that made her feel. This pile-up of daily pain, nausea, and an inability to eat, left her feeling "morbidly depressed" she told me, a term rarely used by non-psychotherapists, but one which captured the destitute emotional state in which she found herself.

Using the energy medicine techniques that I'd learned from Donna Eden, I began to rebalance some basic energy patterns in her body and went after the nausea by holding key acupuncture points with my fingertips. To her astonishment, the nausea began to ease within minutes! When I told her that she could hold these points on herself and get results, she was both delighted and highly motivated to try it. I encouraged her to do this every day with the idea that she could re-train her energy system to flow in more functional ways, thereby changing her body's nausea response. Then I cleared and balanced her energy vortices, paying special attention to those that were situated near areas of pain. She could feel the pain beginning to soften as I gently spun these energy centers above her body.

By the third session, her nausea surfaced only occasionally, and her pain was described as "mild." At this point, I began using principles from Energy Psychology to identify and clear thought patterns that might be impeding a more complete level of healing. She tapped on acupuncture points while simultaneously thinking about her pain and nausea, and found that it deactivated the negative emotions associated with her thoughts. "I feel good!" she said, with a look of surprise on her face. She could tell that the pattern of her symptoms was becoming more intermittent, and she knew that she was moving in the right direction. I explained that while most patients feel relief in the first session, there is a cumulative benefit derived from multiple treatments as the energy patterns are repeatedly rebalanced and the body's healing systems are coaxed back into action. By the following week her nausea was completely gone, her apprehension about eating greatly diminished, and she began to eat small amounts of normal food. The week after that, to her great relief, her feeding tube was removed.

When we are in a great deal of pain and discomfort, it's hard to attend to other matters in our lives, and that was true for Sylvia. Her illness and suffering had derailed her from moving through a healthy grieving process over the recent deaths of her parents, but as the layers of her physical symptoms peeled away, she was able to talk about the sadness that she still felt at losing them. To support her energetically through this process, I worked with her lung meridian, which carries the resonance of grief, and she found herself feeling lighter in spirit. Still missing them, but now with a lighter spirit.

Before we ended her energy medicine sessions, she asked me for help with one more thing. "I'm feeling really scared about my next PET scan, and about finding out whether or not the cancer has returned." To soothe her intense anxiety, I had her create a mental videotape of images, thoughts, and feelings regarding the upcoming scan, and asked her to "run it" while I held places on her head called neurovascular points that are known to send a calming, soothing energy throughout the body. In less than 10 minutes her worry melted away; she felt ready to get the scan, and decided she'd just hope for the best.

Like other patients with a serious illness, Sylvia brought a myriad of physical, emotional, and spiritual concerns to our energy medicine sessions. She began to understand that each stress that we cleared from her bodymind would move her to a deeper level of peace, helping to create a more optimal climate for the functioning of her immune system.

I was elated when I received a phone message that her scans were clear. At her tenth and last session, Sylvia summed up her appraisal of energy medicine by saying, "My nausea is gone, my pain medication went from 60 mg to 5 mg a day, I'm eating normal food, and I feel joy in my life again!" She was amazed that it had all been accomplished without the use of further medication. Two years later, Sylvia called to tell me she was still feeling great.

Pat Butler has seen a marked shift in public attitudes toward energy medicine in the past decade. "Patients used to make comments like, 'I don't know if this fits into my religious beliefs,'" she observes, "but not any more. There is a growing awareness that there is a wide set of possibilities for optimal treatment." This awareness is penetrating the research community as well; a group of University of Illinois, Chicago, researchers recently wrote that, "instead of trying to fit cancer into a theory employing genetic determinism, we have accepted that dependence upon genetic determinism to explain cancer has become obsolete,"[2] and researchers as well as hospitals are more open to exploring how energy medicine and conventional medicine can both benefit patients with cancer and other serious diseases.

Discerning an Appropriate Treatment Path

Andrew Weil, M.D., has developed a simple formulation to allow a patient to determine whether conventional or alternative medicine is right for their particular condition. In his book *Spontaneous Healing*, he lays out a set of criteria that allows both patients and medical professionals to "stream" patients into either alternative or conventional therapies. It's important for each of us to have at least a rough idea of whether our condition can be effectively treated by allopathic medicine, or whether the best leverage point comes from an alternative therapy.

One of the most intriguing experiments in setting up the clinic of the future—one that encompasses both conventional and alternative approaches—is an establishment called the Integrative Medical Clinic of Santa Rosa (California), or IMCSR. Robert Dozor, M.D., and his wife Ellen Barnett, M.D., Ph.D., established their treatment center in 2001. Walking into the clinic, the first thing the visitor sees is a wide hall, with a fountain and beautiful artwork, which then opens up into a large room with comfortable couches, an altar, another fountain, and a holistic health library. On the way, there is a glass-fronted herbal dispensary, and a receptionist's desk.

*Fountains, curves, and open space in the waiting room at the
Integrative Medical Clinic of Santa Rosa (IMCSR)*

*Kwan Yin, the Goddess of compassion, on the altar
in a corner of the waiting room at the IMCSR*

Comfortable chairs, plants, and a patient lending library at IMCSR

Among the practitioners at the ICMSR are a chiropractor (D.C.), a somatic therapist (CMT), an acupuncturist (L.Ac.), a marriage and family therapist (MFT), a naturopathic doctor (N.D.), and a psychologist (Ph.D.). In addition, doctors Dozor and Barnett have a compendious knowledge of Ayurveda, Chinese medicine, herbalism, and other alternative medical specialties.

The treatment protocol is coordinated by a computer system that tracks patient files and updates them in real time. During a consultation with the naturopath, for instance, he enters his notes on a screen that is immediately available to all the other practitioners. He might also walk ten steps down the hall to consult with the psychologist or with the family physician. If a patient goes from him to the physician, the physician has an immediate record of exactly what the naturopath's findings and prescriptions were. This system is essential to the coordination of treatment by so many different specialties.

Navigating the Options

Barbara Marx Hubbard, one of the world's pre-eminent futurists, points out that many of today's occupations did not exist fifty years ago. She writes: "Just as many of the new functions that people do today—biotechnician, telecommunication specialist, nanotechnologist, environmentalist, futurist, medical ethicist, desktop publisher—didn't exist in the 1950s, the job descriptions that reflect our emerging evolutionary vocations or callings are yet to be defined."[3]

One of the most interesting ways in which the IMCSR points to the future is that it has required the invention of a new job: an intake specialist, or "Navigator" in IMCSR parlance. This highly skilled person meets with patients, and assesses which treatment, or combination of treatments, might work best for them. The mix might change as they visit different practitioners within the building, but the Navigator has responsibility for making the initial determination. Early planning for the IMCSR assumed that a physician would have this responsibility, but the insurance compensation system of the current American medical system means that the physicians are the financial drivers of the clinic; it is their time that generates the most revenue and supports the rest. The structure of the clinic was adjusted accordingly, and the job of Navigator created. In the last decade, many other complex medical systems have invented similar jobs.

Genetic therapies benefit all humans; all ethnic groups are essentially the same genetically. In their funny and provocative book *Mean Genes,* Harvard Business School scholar Terry Burnham, Ph.D., and his colleague Jay Phelan, Ph.D., a biology professor at UCLA, tell us that: "Using advanced DNA technology, measures of genetic variation confirm that human races are trivially different from one another. For that one-quarter of our genes for which there is some variability, there is little rhyme or reason to how this variation is divvied up from one person to the next. Africans have huge variations in blood type: some are type O, some AB, others A or B. But the same goes for Asians and Turks, Russians and Spaniards."[4] This means that the psychotechnologies that are emerging from the latest research can be used by health care workers in Uganda or Laos—as well as in Scotland or Canada.

Intention as Quantum Conversation

It is possible to imagine a time in the near future when *paying attention* will be the first thing we do when we get sick. Spiritual and emotional remedies will be the first line of defense, not the last. Sufferers will seek metaphysical solutions *not* when they've *exhausted* all conventional means, but instead *before* they submit to the drugs and surgery of allopathic medicine. Allopathic medicine might become a medicine of last resort, rather than of first. It will be used to treat certain conditions for which it is admirably suited. But others, especially some of the recently-recognized and still-mysterious conditions like autoimmune disorders, will first be treated with non-invasive methods meant to shift the patient's energy field. Dr. Shealy estimates that surgery or drugs are required by no more than 15% of patients.[5] Many, perhaps most, of the times drugs or surgery are used today are inappropriate. Alan Roses, M.D., vice-president of genetics for Glaxo, one of the world's largest drug companies, startled the medical world in 2003 when he frankly admitted that most of the industry's drugs are ineffective. Britain's *Independent* newspaper reported that Roses "said fewer than half of the patients prescribed some of the most expensive drugs actually derived any benefit from them.

"It is an open secret within the drugs industry that most of its products are ineffective in most patients but this is the first time that such a senior drugs boss has gone public. His comments came days after it emerged that the NHS [Britain's universal health service] drug's bill has soared by nearly 50 per cent in three years."[6] Another article, written by respected medical reporter Susan Kelleher for the *Seattle Times,* reported on the disturbing trend that "some of America's most prestigious medical societies take money from the drug companies and then promote the industry's agenda."[7] The result is that many people are taking drugs that may be more of a threat to them than the disease for which they are being treated, or "even kill them."[8]

Money infects medicine to the detriment of patients. A comparison of studies of drugs used to treat schizophrenia found that the pharmaceutical company whose drug came out on top in the study was almost always the company that had paid for the study.[9] And even when they're prescribed, doctors are not always doing patients a favor: "Pharmaceuticals temporarily diminish anxiety, panic, and depression; decrease our emotional and physical pain; or kill hostile germs in the body. Yet they leave the root causes of our ailments untouched and diminish our capacity to feel. And our feelings contain vital information that is essential for our health, well-being, and peak functioning."[10]

The closer you scrutinize the alternatives, the more attractive energy medicine becomes. A PSYCH-K or EFT session—safe, non-invasive, and usually effective—is a much more compelling first line of treatment than conventional medical therapies for most conditions. It shifts our maladaptive beliefs and irrational fears, rather than suppressing them with prescription drugs.

Precise Energy Leverage Points

The energy diagnosis of the future might be focused on discovering the precise leverage point required to produce healing. Imagine a logjam on a river, hundreds of logs piled up producing a blockage through which nothing can pass. Removing the logs one by one is a tedious process requiring the expenditure of great stores of energy.

But if you find the right log or logs that are the linchpins around which the pressure pattern of the whole jam takes form, and remove it, suddenly all the other logs are released too, without the massive application of work required to remove them all one by one. In the same way, each patient responds differently. One might benefit most from aromatherapy, another from acupressure, another from somatic therapy. One person's logjam is another person's forest.

Then within each discipline, good therapists notice the particular logs that are holding up the flow. A good acupuncturist may stimulate only the three or four points required to restore balance to the body's energies. A skilled somatic massage therapist will go straight to the zone that releases the most tension the fastest. "How did you know where to touch me?" amazed patients often ask. Good professionals can quickly spot where the maximum point of pressure is in the mass of logs, and pull out just that one log that releases the rest. Like healing my two negative beliefs about money, when the precise leverage point is discovered, you can focus on healing only those items that are critical in holding you back, the single logs in the logjam that might be inhibiting vitality throughout your system.

The skilled diagnostician and prescriber of the future will be trained to spot the possible constellations of jammed logs, and, by going straight to the point of maximum leverage, offer the patient effective relief. Patients who present with various conditions might be trained to discover which thought, linked to which belief, linked to which strong emotion, will release their particular logjam. When precise energy leverage techniques become the front line of medicine, many conditions from which people today suffer will become a thing of the past.

Medicine, wellness, and healing look very different in a quantum world than they did in the mechanistic and reductionist world that preceded it. Each new discovery is another indication of how important consciousness is for healing. We are learning to see our cells and our bodies as malleable, influenced by every thought and feeling that flows through us. Knowing this, we can choose to take responsibility for the quality of thought and feeling which we host, and choose those that radiate benevolence, goodwill, vibrance, and wellness. Doing this, we positively affect not just our own wellbeing, but that of the entire world of which we are a part—and the great ocean of consciousness in which our individual minds swim.

14

Medicine for the Body Politic

Start by doing what's necessary; then do what's possible; and suddenly you are doing the impossible.

—St. Francis of Assisi

"A woman trained in EFT[1] was at an elegant dinner party when one of the guests began to go into anaphylactic shock. Anaphylactic shock is a rapid and severe allergic reaction to a substance (most often a vaccine or penicillin, shellfish, or insect venom such as a bee sting) to which the person has been sensitized by previous exposure. It can be fatal if emergency treatment, including the administration of epinephrine injections, is not given immediately. Apparently, this man was severely allergic to shellfish and he was unknowingly eating crab-stuffed ravioli. As his face and throat began to swell, the host jumped up to call 911. The woman immediately took the man into another room and began to treat him with EFT. Before her eyes, the swelling in his face and neck began to go down and in just a few minutes he had returned to normal. He rejoined the dinner party and the 911 call was cancelled as all his symptoms completely vanished. All this occurred within a ten-minute period."[2]

Such stories point to the possibility that Energy Psychology can be routinely used by non-professionals for the benefit of those around them, for the social good. Like emotional CPR, they can be used as an emergency medical intervention. If schoolchildren learned these techniques to help them cope with playground stresses, if hospital patients learned them to help them cope with pain, if people in social crises used them to help stabilize their emotions, then panic and powerlessness would no longer thwart sensible action. The reservoir of sane, stable intellectual and emotional power that lies within us would be unlocked and brought to bear on the social problems that lie before us.

Discontinuous Social Change

Just as sudden healing miracles are a possibility for our bodies, sudden social miracles are possible for our culture. We are living at a time of profound social discontinuity. Patterns that have been under the radar for hundreds or thousands of years are coming into the light of healing.

Examples are all around us. For millennia, women were suppressed by society. In most countries, laws prevented them from having social status equal to that of men. Often women were not allowed to own property; often they were treated as property.

In Western societies this has changed—all in the course of just a century. In 1893, New Zealand became the first country to give women the right to vote in national elections. In Britain women won the right to vote in 1918, (1920 in the U.S.), and in the second half of the twentieth century, the women in most of the world's functioning democracies obtained similar rights for themselves. In most Western countries, women are now on a roughly equal social, financial, and political footing with men, at least on paper. This means that the vast creative potential of half of humankind, locked up for centuries, has suddenly become available to society. Women's rights are an example of discontinuous social change.

LEAFLET No. 1.]

WOMEN'S SUFFRAGE.

Are Women Citizens?

Yes! when they are required to pay taxes.
No! when they ask to vote.

Does Law concern Women?

Yes! when they are required to obey it.
No! when they ask to have a voice in the representation of the country.

Is Direct Representation desirable for the interests of the people?

Yes! if the people to be represented are men.
No! if the people to be represented are women.

All who believe that this state of things is neither just towards women nor advantageous to men are invited to become members of "The Victorian Women's Suffrage Society."

PLATFORM.

To obtain the same Political Privileges for Women as are now possessed by Male Voters, with the restriction of an Educational Test by writing legibly the name of the Candidate on the Ballot-paper.

ELIZTH. H. RENNICK,
Hon. Sec. and Treas

"AFRIKA," SHIPLEY STREET,
SOUTH YARRA.

N.B.—Subscriptions received by Miss O'Riordan, Ladies' Club, Collins-street East, and Miss Taylor, 86 Russell-street.

Leaflet printed by the Victorian Women's Suffrage Society

Another example is incest and other forms of child abuse. For most of recorded history, children were entirely under the power of adults, even if they were being dreadfully abused. For centuries, children were routinely employed in mines in Britain, under the most appalling conditions, such as sixteen-hour shifts carrying coal on their backs. Incest was such a taboo that it was not even discussed, and was often presumed not to exist. Suddenly, in the last two centuries, we have witnessed rapid, discontinuous social change. England passed the first child labor laws in 1802, and strengthened them throughout the century. In 1916 U.S. President Wilson pushed a child labor law through congress, only to have it struck down by the Supreme Court. Finally in 1938, the Fair Labor Standards Act came into force. Most Western countries also have movements protesting the importation of goods made with child labor in other countries.

Child "hurriers" pushing a coal tub in nineteenth-century London

Another is addiction. In the slums of London in the 1800s, a stereotype of the workingman was that he got his wages on Friday, went to the pub, drank most of them away, and returned home to beat his wife and children. None of his neighbors thought that any of this behavior was amiss. Probably in Babylon, four thousand years ago, the same man could have received the same nods for the same behavior. Certain social norms have been static for thousands of years.

Today, that same man might be going to an Alcoholics Anonymous meeting, or finding some other way to acknowledge and treat his addiction. Whereas for centuries society tolerated addictive behavior, and had few mechanisms for dealing with any but the most violent addicts, today there are many interventions available. There are twelve-step groups for many conditions—from compulsive shopping, to gambling, to attention deficit disorder (ADD), and for those addicted to nonprescription painkillers. Many individuals have begun to confront addictions that, in past centuries, would have been considered socially normative. Taken collectively, they represent a society that is making a serious dent in its addictions.

Another watchword of today's society is diversity. Today, multinational corporations find ways in which to turn multiculturalism into a benefit, to appreciate the diversity of employees and turn that diversity into a strength. Global companies spend millions of dollars on training and equipping their employees to understand and appreciate differences.

Yet a hundred years before, the corporate giants of the world—think railroad barons, sugar monopolies, the Dutch East India Company, Shell Oil, the Rockefellers and the du Ponts, or Henry Ford's production lines—prized uniformity. The rewards of uniformity date back to the ancient Greeks, who figured out the value of fighting in *phalanxes,* rather than in individual combat, as early as the seventh century BCE. Uniformity of armor and weapons gave these organized groups of soldiers, referred to in Homer's epics, an edge over their enemies fighting singly.

Individuals Lead Society

Society has undergone another about-face when it comes to the expectation that people with cancer will die. I remember hushed conversations from my childhood in the 1950s that went something like this:

Aunt: "The doctor says it's cancer."

Uncle: "How long does he have?"

The only variable, in their minds, was how long the person would live. Cancer equaled death, no ifs, ands, or buts.

It wasn't doctors who began to change that mindset, but patients who refused to die when confronted with terminal diagnoses. There are now hundreds of books, articles, and research studies of exceptional cancer patients: those who, rather than accepting a death sentence, immediately begin taking charge, discovering what they can do to help themselves, and nudging the quantum field, and their cells, in the direction of health.

A similar phenomenon has happened with AIDS. Still officially incurable, there are an increasing number of anecdotes of patients who have later tested HIV-negative, after having been earlier diagnosed with AIDS. In the years after the first AIDS case was diagnosed in the U.S. in June 1981,[3] most AIDS patients died quickly and horribly. Today there are some half-million people alive in the U.S., many in excellent health, who have been diagnosed with AIDS at some point in the previous twenty years.[4] And under the headline "First Case of HIV Cure Reported," in November of 2005, doctors at London's Victoria Clinic announced that conclusive tests had shown that Andrew Stimpson, who had twice been tested positive for HIV infection, was now virus-free. DNA tests were used to confirm that there had been no mix-up in the samples.[5] Society is changing its collective mind about HIV/AIDS being a death sentence, as certainly as it changed its mind about cancer starting a generation ago.

Many of the researchers mentioned in this book have been the target of ridicule, character assassination, or suppression. EFT was banned for a while by the American Psychological Association. EMDR "was regarded as snake oil for years, until a large enough body of clinical studies had proven its effectiveness."[6] Just as, in the eighteenth century, Ignatz Semmelweiss, a Viennese doctor, was ridiculed for suggesting that surgeons wash their hands before operations, or Edward Lister was pilloried for cleaning his hospital with carbolic acid disinfectant, the pioneers of the new medicine and new psychology have faced the entrenched opposition of their professions. Yet because these therapies work—and are often far more effective than anything conventional medicine has to offer—patients are voting with their feet and their dollars, dragging research along behind them, and creating social mind change on the way.

Epigenetic Influence and Wars Between Nations

A critical finding of gene methylation and acetylization studies is that once nurturing is done by one generation, and those genes are expressed, then nurturing behavior, and its accompanying gene expression, is passed to the next generation. Nurturing parents produce nurturing offspring, and neglectful parents produce neglectful offspring.

What would happen if an entire population became less fearful and anxious, generation by generation? Would this result in less aggression, both within the society, and toward that society's neighbors?

Social scientists have found evidence of precisely these effects when studying conflicts between nations. Robin Grille, a psychologist from New Zealand, has done extensive surveys of the parenting literature of warlike populations.[7] For instance, the most popular Prussian parenting manuals in the 19th century stressed that children should be absolutely obedient, and that the most minor infraction should be punished with beating. Beating children several times a day was not uncommon, and was not considered unusual or cruel. It was simply the accepted German nineteenth century paradigm of good parenting.

But it produced the generations who fought the Franco-Prussian war (successfully for the Germans), and then World Wars I and II (unsuccessfully for the Germans). Grille claims that there are many other historical examples of societies that practiced mass childhood cruelty subsequently producing generations of brutal and martially-minded adults.

He also believes that the reverse is true, that societies that introduced enlightened elements of child nurturing became less violent and martial in subsequent generations, and more prone to direct their energies into a quest for justice and equality. This led to social innovations.

In the American colonies in the eighteenth century, children were often raised to be independent and outspoken. Visitors to the colonies remarked on finding confident children who were not afraid to speak their minds. These were the mothers and fathers of the signers of the Declaration of Independence and the authors of the Bill of Rights.

The nurturing of children has other intergenerational social consequences. The author of the best-sellling book Freakonomics proposes a novel and provocative explanation for the dramatic decrease in violent crime in America from the early 1990s onward. The roots of this phenomenon long baffled social analysts. How could crime in a country drop by such a huge percentage in such a short time? Explanations like better policing or longer incarceration are inadequate to explain such a marked shift.

The answer comes from an unlikely source: The establishment of legal abortion on demand. When unwanted children could be aborted, many accidental or unplanned pregnancies were ended this way. Women who were economically or psychologically unable or unwilling to raise their babies had the option of abortion. Many exercised their new legal right.

This meant that most of the babies born were wanted by the mother, father, or both. These children presumably received better nurturing than the unwanted children of the pre-abortion era. The availability of abortion created an abrupt decrease in the number of neglected babies, and an abrupt increase in the number of nurtured babies. We now know that nurturing initiates epigenetic signals that produce less fearful and anxious adults. Twenty years after the legalization of adoption, when both groups had grown up, the pool of violent adults had shrunk, leading to the sudden and precipitate drop in the crime rate.[8] Without knowing it, society, by legalizing abortion, made an epigenetic intervention that resulted in less violence a generation later. Nurtured children produce a more peaceful society.

The Epigenetic Social Cycle

One of the best-known proponents of early childhood nurturing is Michel Odent, M.D., a legendary French obstetrician. Odent groups together the child's time in the womb, birth, and the first year into what he calls the "primal" period, which he believes is formative of many key adult attributes. He says that when "researchers explore behavior, a personality trait or a disease that can be interpreted as an 'impaired ability to love,' they always detect risk factors in the perinatal period." He cites studies that show that, "the main risk factor for being a violent criminal at age 18 was the association of birth complications, together with early separation from or rejection by the mother.[9] ... From studies among mammals as diverse as rats, hamsters, sheep, goats and monkeys one can conclude that there is always, immediately after birth, a short 'sensitive' period which will never be repeated, and which is critical to mother-baby attachment and subsequent development."[10]

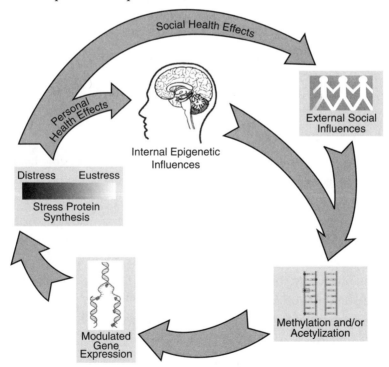

The epigenetic social cycle

A chain of social events which I call the epigenetic social cycle is discernable here. Nurturing of children produces beneficial gene changes, which produces increased nurturing of subsequent generations, and a less violent and safer society. This in turn leads to the synthesis of fewer stress proteins in individuals, and greater production of cell repair hormones such as DHEA. Less stressed individuals are in turn more likely to contribute to social peace.

The same epigenetic social cycle can also lead to a descending spiral of fear, anxiety and violence. The failure of a society to nurture its children creates an epigenetic intervention that suppress their ability to handle stress. They are more fearful and anxious, and more prone to offensive or defensive violence. A violent society is unconducive to childhood nurturing, leading to further epigenetic gene modulation and even greater stress. This leads to increased violence.

These cycles are by no means evident to the fields of medicine and psychology at the moment. Epigenetics is too new for its impact to be assimilated and understood. Epigenetics explains the success of interventions that look like magic to the existing medical and psychotherapeutic models, and is likely to take several years, or even decades, to penetrate into daily practice, even though the social benefits of positive interventions is clear.

The Reaction of the Old Guard

It will be fascinating to see how conventional medicine handles the challenges posed by energy medicine and Energy Psychology. Some of the key players in the current system have a lot to gain by embracing it, while others have a lot to lose. For instance, HMOs, hospitals, and doctors being paid fixed rates for the number of patients they see (called "capitation" in the jargon of the profession) benefit financially the healthier a patient is. Average amount spent per person for "health" care was $6,683 in the U.S. in 2005.[11] A provider who is receiving $6,683 from a patient who is so healthy that he or she makes one or no doctor's visits in a year, and uses only a few complementary and alternative therapies to assist in maintaining peak health, ends the year with a big profit. A provider dealing with a chronically ill patient, on the other hand, can spend more than $6,683 in a single week of treatment, and a patient rushed into intensive care can cost more than $6,683 in a single day.

This figure is projected to rise to $12,320 per person in 2015. That's one reason why Kaiser Permanente, one of the largest HMOs, much-studied for its efficiency, now offers Qigong classes, meditation classes, chiropractic, acupuncture, addiction recovery classes, and other similar therapies. Some Kaiser practitioners also use Energy Psychology techniques. It's not just good health care; it's good business. Such organizations are net winners when people are healthy.

Others are net losers. Psychotherapists with patients who have been seeing them weekly for many years, working on life issues, at an average visit cost of $100, lose an ongoing revenue stream of $5,000 a year when the patient might heal an issue with Energy Psychology in a session or two. Such therapies pose a financial threat. They also suffer from a credibility gap. A therapist may rightly be skeptical that a patient's chronic agoraphobia, for example, can be cured in an hour, when years of therapy have produced only modest gains.

Long-term, however, therapists often report financial benefits from using Energy Psychology, quite aside from the professional satisfaction of seeing patients get better. "The financial benefit comes in being able to charge more for sessions, and overcoming the likelihood that people avoid therapy because it takes too long," says Fred Gallo, Ph.D., the inventor of the term Energy Psychology, and the author of several books on the subject.[12]

When the healing professions see their revenue streams threatened, they sometimes act in ways that protect their pocketbooks but hurt their patients. In the early 1900s in New York, Mary Mallon, an immigrant from County Tyrone, Ireland, later to be known as Typhoid Mary, infected dozens of people with typhoid fever, several of whom died. Though she was herself healthy, she passed the infection along in the course of her work as a cook for wealthy families, often through her favorite dish of iced peaches. Typhoid fever, caused

Typhoid Mary, from a contemporary publication[13]

by a strain of *salmonella* bacteria, is spread by poor hygiene, and in Mary's case, denial, since she believed she was healthy. She maintained her innocence even when doctors confronted her with evidence to the contrary—and to be fair, there was no scientific consensus at the time that a person could carry a disease without an active infection. Mary strenuously resisted efforts by Dr. Josephine Baker of the New York City Health department to have her removed from circulation. Dr. Baker tried to have her quarantined twice, in one case sitting bodily on "Typhoid Mary" till the police could cart her away.

Dr. Baker also trained mothers to wash their hands, limiting the spread of the epidemic. She also insisted that older siblings, who often cared for their younger brothers and sisters, maintain good hygiene, too.

Dr. Baker's success in controlling the spread of typhoid and cholera roused the ire of the traditionalists of her day. Thirty Brooklyn pediatricians petitioned the mayor to pull the plug on her activities, because the supply of sick children being brought into their lucrative practices was drying up. Today, a conventional medical system under threat from many fronts has to confront new therapies that can heal patients much faster—and, in many cases, more cheaply—than conventional approaches. Some react with ridicule, scorn, dismissal, and an unscientific refusal to honestly consider the facts. Yet surfing through the Energy Psychology web sites reveals the names of thousands of physicians, social workers, psychiatrists, and psychologists who have eagerly trained in these new methods for the benefit of their clients and patients.

Energy tapping is to the twenty-first century what hand-washing was to the nineteenth. It's a technique that's so simple, accessible, and low-tech that it doesn't seem possible it can deliver the results it promises. It requires no expensive gadgetry, elaborate explanations, board certification, years of professional training, university degrees, or membership in prestigious societies. It seems preposterous to those with an investment in today's medical system that energy tapping could have such powerful results. Yet clinical practice is proving the worth of this method every day, and science is rapidly providing a base of evidence for the efficacy of this simple intervention.

Clearly, as a society, we should be studying the best examples of healing we can find, even if they seem like quackery or story-telling at first glance, and apply those principles for the benefit of all sick people. And we must sharpen our focus on a paradigm of health that seeks optimal vibrant wellness, instead of one that allows unrecognized low-level malaises to mature into full-blown diseases, which we must then treat—often causing misery to the patient while squandering large sums of money.

Emotional Peace in the Wake of War

Two thousand years ago, Jesus said, "Blessed are the peacemakers." Yet it is only in the last century that we have seen institutions emerge to study and develop the skills to make peace. The League of Nations was a first faltering step on an international scale. The United

273

Nations is far from perfect, yet has catalyzed peace in many regional conflicts that would have been worse without UN intervention.

One of those is Kosovo. Though a province of Serbia, only a minority of the population was Serbian. When the Kosovars became politically restive, the Serbian army went in. Civilians were attacked, and in some cases massacred. The Serbian army used rape "as an instrument of intimidation," as it had earlier done in Bosnia and Croatia. Thousands of Kosovars fled. The Serbians ignored appeals from the European Union and the UN to stop the killing and violence. Only when the United States began a high-altitude bombing campaign, were the Serbs eventually deterred.

Kosovar refugee with deserted village of Milic in background[14]

A volunteer team of Energy Psychology practitioners went to Kosovo in an attempt to help survivors deal with the severe emotional wounds left by the attacks. Their experiences were reported in the *Journal of Clinical Psychology*.[15] They treated 105 people, most with Thought Field Therapy, or TFT. Based on the victims' self-assessments, they experienced complete recovery from "the post-traumatic emotional effects of 247 of the 249 memories of torture, rape, and witnessing the massacre of loved ones" that were treated.

The chief medical officer (surgeon general) of Kosovo wrote the following letter of appreciation:

> Many well-funded relief organizations have treated the posttraumatic stress here in Kosovo. Some of our people had limited improvement but Kosovo had no major change or real hope until...we referred our most difficult patients to [the international treatment team]. The success from TFT was 100% for every patient, and they are still smiling until this day [and, indeed, in formal follow-ups at an average of five months after the treatment, each was free of relapse].[16]

Similar results have been shown in work with Hurricane Katrina survivors, Pakistani earthquake survivors, and those dealing with the effects of other disasters. Charles Figley, founder of Green Cross, which addresses the psychological needs of disaster victims, has said that, "Energy Psychology is among the most powerful interventions available to us."[17]

In June 2006, Sarah Bird and Paul O'Connor, Irish humanitarian volunteers sponsored by the Association for Comprehensive Energy Psychology, traveled to Kashmir to help earthquake victims. They were shocked at what they saw: "Eight months after the earthquake Mussarafabad is still a hellhole for the people who lost everything. They are still living in tents, shanties, and hut-like structures. There was huge devastation here; the mountain swallowed up whole communities and the figure they are talking about unofficially is closer to 200,000 deaths." They worked with schoolchildren, as well as in the hospitals. They had many profound encounters, of which this is typical:

> We went to see the head psychiatrist on our arrival and he had a woman in with him who was seeing him for the first time. She had lost her children and hadn't been able to sleep since the earthquake. She was complaining of severe headaches. When we went into his room she was softly banging her head on his desk. After about five minutes of this I could not watch anymore and I asked him if I could sit with her. I just tapped her hand until she raised her head and then held her head in the TAT pose until she relaxed. The doctor was translating; she said she felt lighter but the head was still bad. So we tapped some more and you could see her visibly relax. The doctor gave her some medicine for the head but also told her to use TAT every day!

Fred Luskin, Ph.D., whose Stanford Forgiveness Project has taught thousands of people how to let go of wounds and resentments, has offered similar trainings in other troubled parts of the world. "Luskin says resolving such resentment 'replaces hostile feelings with positive ones that make your body feel calm and relaxed, which enhances health.' In one of his studies, seventeen adults from Northern Ireland who lost a relative to terrorist violence received a week of forgiveness training. Their mental distress dropped by about 40%, and they saw a 35% dip in headaches, back pain, and insomnia."[18]

EMDR has a Humanitarian Assistance Project which has worked in the Middle East, in New York after September 11, and in many other trouble spots around the globe. The following story comes from the efforts of a team in Palestine:

> In one memorable instance, a Palestinian father of four underwent EMDR as part of [the EMDR Humanitarian Assistance Project]. When he began the session, he was filled with homicidal rage toward all Israelis. At the end of the session, he reported feeling "much better" and spoke these words:

"You must always remember: where there is life, there is hope." In the 18 months since his session, this man has tirelessly worked to establish EMDR programs for children in West Bank refugee camps. He has given permission for his story to be told in order that others might also have hope.[19]

The effects of such improvements in people's lives has great benefit to society as a whole: "A study that tracked the clinical outcomes of 714 patients treated by seven therapists using Thought Field Therapy (TFT) in an HMO setting[20] found that decreased subjective distress following the treatment was far beyond chance with 31 of 31 psychiatric diagnostic categories, including anxiety, major depression, alcohol cravings, and PTSD."[21] Just imagine if every patient had access to such therapies as a routine part of treatment. Savings from the social costs of alcoholism and depression alone would far outweigh the costs of treatment. Ill will between warring factions and groups could be treated before it reaches out to devastate the whole society. And conflicts could be defused before they become conflagrations, with all the resulting human tragedy.

Another organization that uses Energy Psychology and energy medicine interventions to treat disaster victims in developing nations is called Capacitar. Capacitar founder Patricia Mathes Cane, Ph.D., has worked in most Central American nations, including Guatemala, which was ravaged by a 35-year civil war. During the war, there were instances in which the entire population of men and boys in a village was shot, and the bodies buried in mass graves. The degree of psychological trauma of the survivors, mostly women, is hard to overstate. Dr. Cane has found Energy Psychology techniques, especially TFT, to be invaluable in treating such people. She has written several workbooks, in both English and Spanish, to guide relief workers and others in these situations. She believes that Energy Psychology techniques are actually culturally more appropriate for such populations than traditional psychotherapy. In her book *Trauma Healing and Transformation,* she writes, "Often the client-therapist model does not fit the needs of grassroots people, who for most of their lives have been disempowered by state, church, educational and medical institutions."[22] Self-treatment can avoid triggering the power dynamics inherent in the psychotherapy techniques with which most therapists and their clients are familiar.

Women of the Mayan community of Solola, Guatemala, using TFT[23]

Wanted: One Thousand People

Under the impact of changes in consciousness, our social DNA is changing as surely as our physical DNA might do.

The cells of the immune system are few in number compared to the trillions of cells in our bodies. Yet they play a dominant role in keeping the whole organism healthy. Every cell in the body does not have to be an immune system cell in order to fight off infection. It only takes a few.

In the same way, it does not require that all of the individuals in a society wake up in order for the whole of that society wake up. It does require a critical mass, but that mass can be surprisingly small. Remember that the Renaissance, a cultural earthquake that completely transformed medieval society in just twenty-five years, was sparked by *a mere one thousand people.*[24]

Like the few who sparked the Renaissance and changed the world forever, today we are witnessing a new revolution fomented by just a few thousand courageous volunteers who are aware not just of the effect of their own thoughts and feelings on their genetic expression, but of themselves as the change agents for society. By transforming their own consciousness, they are spearheading the transformation of the whole. They model what a transformed person can look like; they are the social genes that activate the immune machinery of an entire society.

Are you one of them? If you are one of the many thousands of people who will purchase and read this book, you have probably already self-selected yourself as one of those social catalysts. To an individual gene, the task of getting a trillion-celled organism to change looks formidable, even though we know that the organism must awaken if it is to survive. Awakening is urgent.

Yet once you realize you're acting in concert with a system, that you are not isolated, that your mind and your experience are subject to quantum entanglement in a multiverse of possibilities, then everything changes. The little effects you have in your daily life become part of one large effect. That effect is large enough to change the world. It already has. And the results of humanity becoming a conscious co-creator in the great ecology of being are beyond our wildest dreams.

15

Ten Principles of Epigenetic Medicine

The winds of grace are always blowing, but you have to raise the sail.
—Ramakrishna

Today, as I write these words, I'm sitting on a bench in a park on one of Northern California's sunny winter days. My laptop and a notepad sit on my lap. I watch as my youngest son, Alexander, plays in the sand with his bucket, spade, trains, and cars. "I'm digging for buried treasure!" he exclaims excitedly, as his toy rake hits a buried object. Then, as he unearths it, he says, "That's garbage!"

A toddler runs up, his drooly smile proudly revealing two brand new teeth. He picks up Alexander's toy train and begins waving the pieces about. Suddenly he discovers that the engine and the tender snap together. He snaps them and unsnaps them over and over again. He's just discovered a major secret to the way the world works.

Meanwhile, Alexander has run to the slide. After a few minutes, he and a diminutive little girl discover that they don't have to slide down the slide in a seated position. They experiment with going down sideways, head first, and on their backs. They have to show me each new trick. First Alexander counts: "One, two, three, twenty-eight, seventeen, nine, hurry, watch me!" and then he slides down.

As I contemplate the discovery of epigenetic control of genes and watch the joyful play in front of me, I realize how much scientific discovery resembles the playful children in the park. We humans discover how things fit together. We experiment with hypotheses, some of which fail. Even after we've made an exciting new discovery, we are not content. We immediately push forward into new territory. When our knowledge crosses one horizon, we quickly push forward to another, with scarcely a backward glance.

Applying the discoveries of epigenetic control of the genes, and hence cellular function, offers great potential for medical breakthroughs. It allows us to seek therapies that intervene at the level of consciousness by changing beliefs and behaviors that interfere with health. Changing energy patterns before they manifest as disease works at a level of cause that is of a higher order than matter. While metaphysicians have advocated such a focus for millennia, modern research is now giving us an understanding of the genetic changes that occur in response to changes in energy and consciousness. This opens up the prospect of using consciousness deliberately, as a planned medical intervention.

Like the toddler who has discovered that the two parts of the train snap together, science has taken the first tiny steps in charting the mechanisms of epigenetic control. Now, a huge new panorama of research presents itself. There are undoubted widespread positive effects that come from belief change and meditation. Yet researchers cannot be content with understanding only that there is a generalized benefit to these practices. They must begin to chart the precise pathways by which such therapies have their effects, so that the power of consciousness can be harnessed with precision.

Epigenetic medicine seeks these precise pathways. It's not enough to prescribe meditation for depressed patients, even though we know that meditation can be useful in alleviating depression. We need to develop a battery of interventions at that we are reasonably certain can be effective, and that are reasonable for the patient to perform. Consciousness-based epigenetic shifts aren't as mechanistic as taking a drug; they require a degree of self-awareness that is not required of the patient popping a pill along with the morning's tooth-brushing routine.

Epigenetic medicine based on consciousness works seamlessly with conventional medicine, alternative medicine, and integrative medicine (medicine that combines the best of conventional and alternative approaches). For instance, molecular *biomarkers*, "signatures of genes or proteins that are specific to a disease,"[1] allow diagnosis of many conditions long before they might show up on an X-ray or manifest as symptoms. Testing for biomarkers is also non-invasive and safe. Biomarkers have the potential to allow early detection of cancer and other diseases, and biomarker-based tests are expected to become widespread in the coming decades. Dr. Leroy Hood, cofounder of Amgen, the world's largest biotechnology company, and in whose laboratory the DNA sequencer originated, forsees a day when "with just a single pin-prick, a nanotechnology device will quickly measure and analyze 1,000 proteins in a droplet of your blood."[2]

DNA screening is now becoming widely used. There are "more than two dozen online genetic testing services springing up to take advantage of advances in genomics.... Tests for nearly 1,300 ailments have been developed so far."[3] Newer tests are being develop that use techniques even more sophisticated than gene chips. DNA testing is joining hands with nanotechnology to produce tests that can determine the expression not just of a gene, a sequence of DNA, but of individual molecules, the "building blocks of gene expression."[4] As genetic testing becomes more sophisticated, consciousness-based epigenetic interventions could be developed to target these conditions. Used in conjunction with known epigenetic visualizations, gene and other biomarker tests could go hand in hand with integrative medicine in a comprehensive treatment plan.

Imagine if each of us constructed a set of visualizations that was uniquely geared to our psychological structure. Imagine going to your preventive medicine physician or psychiatrist and having a caring expert help you identify your unique psychological triggers, levers that can aid you in maintaining health. You walk out of the office visit with a prescription pad of beliefs, concepts, prayers, and visualizations that have been scientifically verified to boost your immune system. You then spend ten minutes a day on this routine. In this way, epigenetic medicine has the potential both to treat disease and help keep you healthy.

We already have an outline of the potential effectiveness of epigenetic medicine in our studies of nurturing in animals and adults, from researching the brains of meditators, by examining the beliefs of sick patients, and all the other ways in which we humans can use our consciousness to affect our wellbeing and happiness. Recognizing that we are conscious energy systems giving rise to matter results in an approach to medicine that is completely different from one that treats the material substance of our bodies as though it were primary. The following list of principles provides guidelines for treatment that spring from the primacy of consciousness. They are the foundation for a medicine that detaches from an obsession with making matter look a certain way, and instead supports the expression of the health that is inherent in consciousness.

There are, I believe, a number of interwoven and mutually reinforcing principles for treatment that can be derived from the emerging consensus of knowledge and experience collected in the infant field of epigenetic medicine. Some of them are:

1. Intentions First, Outcomes Second

The literature of medicine is littered with the word outcomes. The eventual material result, the "outcome," is the end point that studies use to determine the efficacy of a particular treatment. The medicine of consciousness, on the other hand, focuses first on the wellbeing of the soul, and then on the epigenetic effects of that wellbeing on the cells of our bodies. Symptoms are perceived as guides to understanding the needs of the soul, not as nuisances that must be made to disappear.

"Cancer was the best thing that ever happened to me," is a paradoxical refrain that echoes through the annals of survivors' stories. The "outcome" for such a patient may still be cancer, but the inner experience may be healing. Illness can be full of gifts, if we shift our perspective to the present moment and detach from an obsession that our bodies be in a certain state.

The quantum soup has an infinite number of possibilities. Rather than get too attached to any one outcome and cling neurotically to our preferred result, we can instead state our intentions clearly, then let go. Reinhold Niebuhr's serenity prayer used by AA and other twelve-step programs is a powerful mantra of non-attachment: "God, grant me the serenity to accept the things I cannot change; courage to change the things I can; and the wisdom to know the difference." It isn't whether you live or die in the end that counts, it's how healed your life is at any moment. Western medicine, with its focus on outcomes, must shift to realize the truth of the Zen dictum: "The journey is the destination."

In another paradox, it is often by letting go that we gain something. Setting a clear intention in our consciousness is often a better way to get what we want than manipulating the events around us to attempt to produce it. Clear intentions open us up to streams of quantum possibility that our ego-bound imaginations cannot grasp.

2. Healing Is a Process, Not an Event

Modern medicine has an underlying structure that perceives illness as an event. The event starts with "symptoms," and ends with a "cure." Center stage in the event is the prescribing of a drug or performing of surgery. But most of health is not an event. It is a process. It is not a rock, but rather a river. Lifestyle changes such as diet and exercise can address systemic problems, but involve a change in life-process day after day after day. Weight loss isn't being twenty pounds lighter (an event); weight loss is what I eat and how I exercise in the next few hours (a process).

Joining a gym is an event, and an even-centered perception of health encourages people to do this. Many people join gyms every year. But "up to 65% of new gym members drop out in the first six months,"[5] and more thereafter, as a result of seeing health as an event rather than a process. A process is something we incorporate into our lifestyle, like having a habit of going to the gym every Monday, Wednesday, and Friday, and completing a set routine.

The same applies to spirituality. At last Easter's sermon, attendance at my church was dramatically greater than usual, as is typical for holidays. Starting his sermon, the minister laughed and said, "Now I'm going to talk to all of you that I last talked to last Easter, and won't see again till next year."

Going to church on holidays is a symptom of seeing worship as an event. A daily prayer practice is a process that enfolds every event of the day. It is in these regular daily choices that new quantum potentials are activated.

Researcher William Collinge has a useful visual aid to conceptualize the healing process. He represents it as wavy line with a series of ups and downs. There is a turning point, imperceptible at the time, that marks a shift to higher highs and an upward trend.

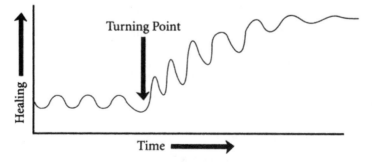

The healing pattern[6]

The Kaiser Permanente medical system now offers drug and alcohol counseling as part of its services to members; its trustees understand that if patients master their addictions, the later medical costs and consequences will be much less. Shifting to a process view of healing, with a daily flow of healthy choices, lowers the probability of catastrophic medical events.

3. Heart-Centered

Are you in love with your doctor? Is your doctor in love with you? Do you feel a warm glow whenever you think of your doctor? Do you believe that your doctor feels kind and warm feelings towards you? A heart-centered connection (though not a romantic one) is an essential aspect of the healing encounter.

The quality of interaction can be as important, or more important, than the content of interaction. Simply being met by another human being, heart to heart, at the level of soul and emotion, can be a profoundly healing experience. This is the state that people in love find themselves in, and it triggers a cascade of powerful hormonal responses. If every medical encounter began with a heart-connection, and did not proceed until that connection had been established, the content of those encounters might be much more powerful. In a landmark paper by Andrew Weil, M.D., and Ralph Snyderman, M.D., published in the *Archives of Internal Medicine* in 2002, the authors conclude with a list of six reforms that build on the platform of sound science, yet also focus broadly on the wellbeing of patients. They urge "far more meaningful patient-physician relationships...[that]...provide compassion, [and] provide close attention to our patient's spiritual and emotional needs..."[7]

4. Being-Focused

People's state of being might have as much to do with their health as the puncture wound in their arm. Larry Dossey, M.D., recounts the story of doing his rounds in a coronary care unit and asking patients (all men) why they were there. They had seemingly succumbed to sudden and unforeseen heart attacks. But the majority of answers they gave were based in their life-situations, not their medical histories. Typical responses were, "I couldn't stand to see my boss's face one more day" or, "I feel trapped in my marriage. I can't abide being with my wife" or, "My kids fight constantly. I would do anything to get away from their constant bickering." Their heart attacks had resulted in them getting away from the conditions that were intolerable to them—at the cost of their health. Ongoing focus on the quality of a person's being might prevent them having to produce extreme symptoms to catalyze a change in their lives.

Our presenting symptoms may have a great deal more to do with our state of being than with our medical histories. They may hold keys to our wellness that can make or break our medical histories. To fix the medical problem, while leaving the soul unaddressed, at best defers the consequence.

Some practices—a rich social network, consistent spiritual practice, an authentic vocation, the ability to speak one's feelings, meditation—have been shown by research to build a more powerful sense of personal wellbeing. The physician of the future might first look for the practices that can most bolster the patient's soul, like James Dillard writing on his prescription pad, "Long talks with your Rabbi," before even starting work on the presenting condition. Once the being is creatively expressed and at peace, we can see what symptoms—if any—remain.

5. Treat Whole Systems

When we affect any part of an energy system, we affect the whole. The benefits of complementary and alternative medicine that treat the whole person are becoming more apparent with each new scientific study. We might focus only on the particular log that can unclog the logjam, but we never make the mistake of thinking that a particular log is all that needs treatment. Every particular treatment must be considered in the context of how it affects the whole, and have the goal of improving the function of the whole.

It's easy for medical specialists to see only a diseased uterus and want to remove it, without taking into account the effects on that patient's entire life. "When all you have is a hammer," the saying goes, "every problem looks like a nail." Quantum medicine approaches every symptom as an expression occurring within an integrated energy system, and finds the leverage points that bring that whole energy system back into balance.

6. Healing Before Disease

Energy medicine is not only the place to start treatment of a presenting condition; it is the place to start before there is a presenting condition. A person whose energy systems have been optimized and are functioning well has established a baseline that makes it much harder for disease to take root. Whatever the conditions of our lives, no matter what difficulties we are experiencing in our health, our relationships, our work, the techniques found in energy medicine can optimize entire systems in our bodies, minds, and emotional realms. Because it breaks from a mechanistic model of cause and effect, energy medicine opens our minds, hearts, and bodies to the possibilities of quantum change described earlier. Experiments tell us that healing is not localized in time or space; so we can pray for our own childhoods, we can pray for distant people, and we can pray for the wellness of our planet. A daily healing prayer may be the most effective and benevolent way to set up a base line of healing.

Once we get in the habit of attuning to our inner state of health, we may notice energy disturbances before they manifest as disease. A daily energy self-scan can give us valuable information about how we're feeling that day. Our focus eventually shifts from looking for disease, to seeking ways of raising the bar on the level of wellness we embody.

7. Magnify the Body's Inherent Self-Healing Powers

Sometimes a person needs only a nudge in the right direction to get unstuck from a recurrent pattern and initiate the process of the body's restoration of homeostasis. The first thing a wellness counselor can do is look for those leverage points that might help the process get going. So rather than first looking for outside interventions, the guide of the future will look for the interventions inherent in the patient that the patient might not have seen or might not be utilizing. One of the reforms advocated by Weil and Snyderman is to "Involve the patient as an active partner in his or her care, with an emphasis on teaching each patient the best way to improve his or her health."[8]

A doctor is often with the patient one or two hours a year. The patient is with the patient the other 8,758 hours. Who do you suppose has the most influence on the patient's wellness on a daily basis? Recognizing the enormous healing powers of the body—and finding ways to engage them—presupposes an entirely different model from the classic image of the patient being fixed by a doctor or hospital. A patient accustomed to allopathic medicine might be baffled by an acupuncturist who inserts a needle far from the site of the symptom. Many alternative therapies look for the one log that is producing the jam, and shift just that one. Once it is shifted, the rest of the jam takes care of itself and the body's full power comes to bear on re-creating homeostasis.

8. Stream to Appropriate Treatment Paths

Some conditions are obvious candidates for conventional medical treatment. Others are unlikely to respond to this approach. Trying to treat Chronic Fatigue Syndrome with allopathic medicine is misguided; trying to treat a gunshot wound with alternative medicine is foolhardy. In *Spontaneous Healing*, Andrew Weil offers this simple advice: "Do not seek help from a conventional doctor for a condition that conventional medicine cannot treat, and do not rely on alternative provider for a condition that conventional medicine can manage well."[9] He makes the following distinctions in "what allopathic medicine can and cannot do for you:

CAN:

Manage trauma better than any other system of medicine.

Diagnose and treat many medical and surgical emergencies.

Treat acute bacterial infections with antibiotics.

Treat some parasitic and fungal infections.

Prevent many infectious diseases by immunization.

Diagnose complex medical problems.

Replace damaged hips and knees.

Get good results with cosmetic and reconstructive surgery.

Diagnose and correct hormonal deficiencies.

CANNOT:

Treat viral infections.

Cure most chronic degenerative diseases.

Effectively manage most kinds of mental illness.

Cure most forms of allergy or autoimmune disease.

Effectively manage psychosomatic illness.

Cure most forms of cancer."[10]

This list needs updating, as more and more conditions are being moved from the allopathic column to the complementary and alternative medicine (CAM) column as better research is published. For instance, conventional hormone replacement therapies have been shown to have negative side effects in the ten years since Weil penned this list (the National Women's Health Network calls hormone replacement "a triumph of marketing over science"[11]), while alternative medicine, through exercise and diet-based approaches, plus supplementation if necessary, have been shown to stimulate the body's hormonal production and balance. Electromagnetic stimulation, for instance, has been shown by Norman Shealy to boost the production of DHEA, the most common hormone in the body, and a marker for our degree of stress.[12] Shealy also draws attention to a category that he calls "semi-orphan diseases," those for which conventional medicine is only partially effective. Among the conditions he lists are rheumatoid arthritis, lupus, multiple sclerosis, and chronic hepatitis.[13]

If a provider and patient have these kinds of clear distinctions in mind, it becomes possible to seek appropriate treatment and avoid wasting time, money, and effort on inappropriate treatment. The new category of specialist, like the "Navigator" at the Integrative Medical Clinic, is trained in helping patients (and practitioners) understand these distinctions. Such navigators could become a routine part of the beginning of any treatment plan.

Many holistic practitioners get patients who have not been helped by conventional medicine. Many have been chopped by surgery and debilitated by prolonged prescription drug use. Most of the chronic pain patients who show up at Dr. Dillard's practice or Dr. Dozor's integrative clinic have already gone through the medical mill, with no relief. I can see this changing in a few years, as patients become more aware of the benefits of alternative medicine. Holistic treatment might be the first option they choose, not the last.

9. Revisioning Death

One of the great services that authors like Bernie Siegel, M.D., Elizabeth Kübler-Ross, M.D., and Stephen Levine have done is to shake up the idea, so prevalent in our medical institutions, that death indicates a failure. I remember talking to a grief counselor many years ago about the exciting ideas in Bernie Siegel's book, *Love, Medicine and Miracles.*[14] At the time he was consulting with a high school, doing grief work with students after a rash of suicides. "It is terrible for a client," he objected, "if they try all that touchy-feely stuff and it doesn't work."

Underlying his objection was the assumption that living meant that the "touchy-feely stuff" worked, and dying meant that it did not. Siegel and others have reintroduced into public consciousness the idea of a healed death.

My mother showed me this first-hand. She developed cancer in her left eye and her liver. She had some conventional medical treatments like radiation (chemotherapy was not indicated for her particular condition) and also tried alternative therapies like shark cartilage.

She fit right into image of a person whose life was often in chaos, usually self-created. That chaos regularly spilled over to negatively affect the lives of the people around her.

In the two years before she died, she sought to make amends. She traveled to visit many of the people she had grown up with, and, in person, asked for forgiveness. Her father was still living, and she traveled half-way across the world to see him. She visited her sisters and her childhood friends. Layer by layer, all the heavy weights of a lifetime of resentment and anger dropped off her shoulders. In one of the last conversations I had with her, she agonized over a person she could not locate. She said, "There was a girl in my high school whose name was Helen Freund. I hated her, and she hated me. I've tried to track her down so I can say I'm sorry, but I can't find out where she lives now." She started to cry. I sat her on my knee, and said gently to her, "Mom, I think it's okay if you can't find Helen Freund. I'm sure she's forgiven you for whatever happened, and I know you've forgiven her."

"Doc, I've tried everything! Acupuncture, herbs, EFT, Reiki, prayer, yoga…Now I'm ready for drugs and surgery!"

My mother's heart and soul recovered, but her body did not. She eventually died. But she died at peace, and—in every way that mattered—she died healed. She and my father lived in my sister's spacious home for those last two years, surrounded by friends and family, and she died in the bed where she had slept much of her last few years. It was very early morning when she died, before dawn, but she started up just before the end. Her last words were, "I see the light. Do you see the light?"

When she died, she was honored by hundreds of people. The atmosphere leading up to her funeral was filled with grief, and the rest of the family had decided on an open casket affair, which hardly added to the sense of cheer. So to emphasize the joy she'd come to find in life, instead of the fact of her absence, I had a large-screen TV set up next to the casket. On it, I played a continuous loop of video of her I'd taken a couple of years earlier. In the video, she was telling jokes, laughing uproariously, and waving her hands around to illustrate her points. I thought that the video was essential in order to balance the fact of her death with the magic of her life, and, as well as the tragic story of a lifeless corpse, present the vital spirit of a life fully lived.

The doctor of the future may not say, in hushed tones, "I lost a patient." The patient's ego and body might have died, and a medical ego that sees death as the enemy might indeed see a reflection of its own death in the death of another. But when, as Larry Dossey emphasizes, we understand the survival of consciousness beyond death, we see a change in the form, not in the spirit, and we can celebrate the continuation of that spirit even as we mourn the loss of a form.

10. Understand the Global Context of Healing

Fantastically healthy people cannot thrive on a dying planet. As a society, we have to wake up to the ways in which our personal health fits into the picture of global healing, and vice versa. This will lead to sustainable, rational approaches to health care, rather than medical systems that pay no attention to the waste and cost they incur.

In addition to the financial waste, burgeoning bureaucracies, and unneeded medical tests, procedures, and prescriptions, our landfills are overflowing in part due to the huge volumes of medical waste generated by the large number of disposable objects used in treatment and diagnosis (and their sterile packaging). Their bio-toxicity is also an added hazard to our water supplies and to those who handle this waste.

Fortunately, the field of Ecologically Sustainable Medicine (ESM) has developed, complete with a journal.[15] ESM advances medicine with environmental integrity by offering affordable and renewable medical choices—saving resources and money—while preserving the health of the environment.

Among its goals, ESM advocates the emphasis of wellness in medical practice, the choice of ESM treatments as a first resort, awareness of the environmental impact of medicine, recognition of the importance of ecological health in medical ethics, and awareness of the psychological and cultural benefits of sustainable medicine.

The practice of sustainable medical care necessitates fundamental changes in the delivery of medicine. Whether practicing family medicine, oncology, chiropractic, acupuncture, massage, psychotherapy, or any other medical technique, providers emphasize prevention, precaution, efficacy, and wellness in their daily practice. As a result, the delivery of medicine becomes increasingly more sustainable.

16

Practices of Epigenetic Medicine

As human beings our greatness lies not so much in being able to re-make the world, as in being able to re-make ourselves.

—Mahatma Gandhi

Epigenetics bears promises far beyond our current field of vision. The journal *Science* defined epigenetics as, "the study of heritable changes in gene function that occur without a change in the DNA sequence" That's a good definition of epigenetics as applied to the transmission of information across generations, once seen as the primary function of genes. But it misses the potential of the positive genetic shifts we can create *right now,* in our own bodies, by deliberately making changes in our consciousness that have epigenetic effects.

As I have read the research in this field, it has challenged me in my own life and behavior. There are days when I feel unaccountably grouchy. I know that if I spread my misery into the world around me, I will do myself and others no good. So I make a conscious choice to say and do certain things. I might say an affirmation and do an EFT routine. I might meditate for a few minutes, enter a state of heart coherence, or say a prayer. I might take a ten minute nap. I might force myself to say a kind thing to another shopper in line at the grocery store. I might visualize my work life a year in the future, and see myself looking back on the solutions I found to the problems that so vex me today.

These are all deliberate conscious interventions. My grocery store conversation engages the healing power of altruism. My nap gets my feelings back in synch with my clock genes. An affirmation address potentially harmful beliefs. Tapping affects the electromagnetic field of my body. Saying positive words reminds me that I have a choice to go either way. My visualization reduces my stress, reminding me that, "This too shall pass." Each of these things requires only a few minutes. Taken together, they can turn a bad day into a good day, and give me an experience of peace of mind and health of body. Knowing the power of epigenetic control makes one much less casual about words, thoughts, and actions.

Genetic research seeks beneficial effects by manipulating the composition of genes in the laboratory. Imagine a medicine of the future where the laboratory in which your genes are being modified is your own mind—moment-by-moment, with every thought and every action you undertake. Imagine a virtuous cycle in which heart-focused intention produces a beneficial DNA change, which reinforces heart-focused intention, which accelerates DNA change. Where does the cycle stop? No one knows. In *Life Beyond 100,* Norman Shealy speculates that a human health span, using only the factors known today and not the fruits of future research, might extend to 140 years.[1] This figure seems completely outside the bounds of possibility to most medical practitioners today—perhaps as improbable as proposing to a Utah frontier surgeon in 1900 that, within a century, human life expectancy would almost double from the 42-year average that prevailed at the time....

The idea that our DNA can be reshaped by our feelings, thoughts, and intentions acting in our energy fields might be as axiomatic to the next generation of treatment professionals as today's understanding that aspirin thins the blood. As this idea is researched and developed, an entirely new medicine might take shape. This medicine will be completely different from today's medicine. Taken together with the many other discoveries of the efficacy of complementary and alternative medicine, and the advances in technological medicine, it will shift personal and social wellbeing to an extent we can barely imagine today.

A typical medical visit today goes something like this: A patient makes an appointment, driven to the practitioner by some ailment or complaint. The doctor listens, asks questions, performs an examination, gives advice, and writes a prescription for that ailment.

If the ailment does not go away, or if it disappears but resurfaces in some other form, then further steps may be taken. Tests may be performed. Surgery or more powerful drugs may be prescribed. An escalation of treatment occurs, until the patient "responds."

The first doctor visit is relatively quick and cheap. By the time treatment escalates, for instance into chemotherapy and radiation for cancer, or a hip replacement, the solutions are neither quick nor cheap—and they may have severe consequences for the quality of life of the patient.

This is "back-loaded" treatment, with few of the costs on the front end of the treatment cycle, and a very high total cost in terms both of dollars and quality of life.

An integrative medical approach is quite different. There are more costs, and more attention, on the front end. During the first visit, attention is given to all aspects of the patient, to see how the presenting condition fits into the larger picture. Those larger picture issues are then addressed. If the patient's lifestyle can be shifted, perhaps given nudges by a number of different healing modalities, then many of the medical problems that characterize poor ongoing lifestyle choices may be avoided. The diabetic, for instance, who embraces a diet designed for insulin balance, plus an exercise program, may not need all the costly later interventions that would have resulted from traditional medical treatment. The attention and costs of appropriate holistic intervention result in a much higher quality of life for the patient—and much lower costs over the whole treatment cycle. Here is what such a treatment plan might look like:

1. Start Treatment with Energy Medicine

Energy medicine functions at the levels of the most basic building blocks of consciousness. As such, it is the place to *start* treatment, not a place to go once the remedies offered by allopathic medicine have been exhausted, as so many patients do. Energy systems underlie cellular architecture; they are the first place to start building a foundation for vibrant health.

If you have a serious or life-threatening medical condition, immediate treatment is warranted. "When you've been hit with a poisoned arrow," advised the Buddha, "take the arrow out before you have a metaphysical discussion." The American College of Emergency Physicians offers the following guidelines for recognizing a medical emergency that requires immediate care:

- difficulty breathing or shortness of breath
- chest pain or pressure lasting 2 minutes or more
- fainting
- sudden dizziness or weakness
- changes in vision
- confusion or disorientation
- any sudden or severe pain
- uncontrolled bleeding
- severe or persistent vomiting or diarrhea
- coughing or vomiting blood
- suicidal or homicidal feelings[2]

But most visits to hospitals are not for any of the above conditions. In more than 80% of instances, no identifiable organic ailment can be found.[3] Deciding what constitutes a condition that requires treatment, and for which conditions treatment can be deferred for a time, requires information and self-awareness. Early treatment for certain conditions, such as prostate cancer, makes little difference to the outcome.[4]

Low back pain is another common condition for which a variety of safe treatments based on exercise and consciousness are available. It is pervasive, afflicting some 65 million Americans, "affecting nearly 80% of the adult population at some time during their life. It represents the single most common cause for disability in persons under age 45."[5] Studies show that the practice of hatha yoga by back pain sufferers can reduce "pain intensity, reliance on pain medication, and disability."[6] Yoga has also been shown to reduce the stress level of cancer patients, ease insomnia, and improve emotional wellbeing.[7] It's the first place to start when dealing with low back pain, rather than jumping to drugs, which may mask the symptoms without doing anything to help the conditions giving rise to the pain.

The medical encounter of the future might start with a prayer as certainly as the medical encounter of today starts with a clipboard and a white coat. It might involve Energy Psychology, meditation training, or another epigenetic intervention. Only once safe and non-invasive techniques had been exhausted would risky and dangerous medical interventions be considered.

2. Attachment Doctoring

Without some extra effort, we simply cannot relate effectively to other human beings at the fast pace of modern life. For the gifts of deep human connection to appear, we have to set aside our busy lives and slow down—at least for a time—in order to engage in the meaningful relationships that Weil and Dossey advocate. Andrea Bialek, M.D., whose business card says "holistic gynecologist," and who specializes in menopause, has a busy practice assisting women looking for alternatives to hormone replacement therapy. She says, "A woman will say she has no sex drive, and I may suggest she and her husband go to a bed and breakfast and see what happens when they're happy, healthy, and relaxed."[8] That lovemaking improves on vacation is a truism among marriage and family therapists. When a couple has no agenda for two or three days, no telephones, obligations, or children to deal with, and each person slows down to a pace where they can listen carefully, and speak meaningfully, then gradually match the pace of their mate, a new level of healing becomes possible.

Most medical settings today allow the doctor very limited time with the patient. In the wise words of the authors of *A General Theory of Love:* "Medicine lost sight of this truth: attachment is physiology. Good physicians have always known that the relationship heals. Indeed, good doctors existed before any modern therapeutic instruments did, in the centuries when the only prescriptions were the philters deriving their potency from metaphoric allusion to the healer's own person. The extraordinary results of the lab tests and procedures, the mastery they provided over the wily enemy of disease, proved seductive. Western medicine embraced the effective machines and ceded its historic soul."[9] Establishing a relationship, and *setting a human pace* for the healing encounter, reclaims this historic soul.

One day Angela, my twelve-year-old daughter, was trying to change the diaper of my three-year-old son, Alexander. "Lie down," she commanded. Then, to me, "The baby won't lie down. He never does anything I tell him to."

"Darling," I told her, "Let's tell him slowly." I got a soft cloth out of a closet and laid it out on the counter, while he watched. Then I said to Alexander, "Lie down, baby." I watched the meaning of my words gradually sink in, by observing the slow spread of comprehension across his face. When he eventually grasped the meaning of my communication, he lay down on the cloth, ready for a diaper change. Matching our pace to that of another, especially when we're living fast-paced lives, can take some effort. It is one of the jobs of a parent to notice the pace at which a child comprehends, and match that pace. This is one of the disciplines that can make child-rearing a joy for busy people; it forces them to downshift to second or first gear. It requires paying primary attention to the physical connection with the person you're with, and is part of what prenatal psychologists call "attachment parenting."

The same is true for the healing encounter. The client in front of the doctor doesn't want to rattle off a list of symptoms for evaluation, and is not assessing the doctor by how fast she comes up with a prescriptive solution. The client is suffering, and needs understanding and empathy. Tuning in to people in order to understand their afflictions cannot be done at the same pace as drag racing. This "attachment doctoring" is one that establishes relationship first. Alternative medicine is good at this. Again, from *A General Theory of Love:* "The 'alternative' healers proliferated in response to the demand for a context of relatedness. These limbically-wiser settings are friendlier to emotional needs—they involve regular contact with someone who participates in close listening, and often, the ancient reassurance of laying on hands. Alternative medicine sees these activities as quintessential rather than incidental to healing."[10] A paper recently published by a Johns Hopkins University researcher, Bruce Barrett, summed up a survey of the remarkable internal pharmacy we have at our disposal with the following eight recommendations that caregivers can use to engage it: "speak positively about treatments, provide encouragement, develop trust, provide reassurance, support relationships, respect uniqueness, explore values, and create ceremony. These clinical actions can empower patients to seek greater health and may provide a healthful sense of being cared for."[11]

The healing encounter of the future will be done at the pace of the Integrative Medical Clinic's Navigator—a wise and compassionate ear who can hear the patient's state of being, and then steer him or her in an appropriate direction—not at the pace of the HMO physician, who might have just twelve minutes per patient.[12] The actual therapy session might be quick; I'm happy going to a naturopath who listens to my symptoms and writes a prescription, and with a chiropractor who lays me on the table and makes four adjustments in as many minutes. But the initial encounter has to be conducted at a pace that allows the evaluator to tune into the patient deeply and notice what's happening on every level with that person.

3. Scaled Interventions

In a *scaled* application of treatments, the most benevolent and least invasive therapy is used first. This approach supposes an escalation of interventions, using the simplest ones as the first line of treatment, and employing more drastic means only if the previous treatment is not effective—and using technological medicine only when and if it becomes absolutely necessary. If a patient learns to meditate, begins an adequate exercise program, and makes appropriate dietary shifts, many illnesses take care of themselves. Norman Shealy lists

"I'd prefer alternative medicine."

a number of conditions in his books where innocuous, small-scale changes can reverse the course of diseases that are difficult, costly, and life-disruptive to treat with conventional means.[13] In the late 1970s, I was acquainted with an eccentric old doctor, Henry Wasserman, a professor at New York University whose area of interest was medical ethics. He was horrified at what he discovered in his profession. The most passionate thing he ever said to me in his raspy, cynical, Yiddish-accented voice was, "I have learned enough in this job to give you one piece of solid advice: Never go near a hospital unless you are near the point of death." While he may have had a jaundiced view of doctoring, it's clear that all of us have huge leverage available to us through the power of our awareness, and that consciousness is the first place to go for healing.

Marc Micozzi, M.D., Ph.D., co-editor of the anthology *Consciousness and Healing,* refers to a continuum of approaches, from the most invasive, to the least invasive.[14] A continuum approach is also used by the National Pain Foundation. They advise patients to use a "Pain Treatment Continuum." This is a step-by-step plan "for the logical use of pain treatments that suggests using less invasive, less costly therapies before resorting to more invasive and more costly therapies."[15] They strongly advise relaxation, psychological awareness, and healing behaviors before resorting to more invasive therapies.

Here's an example of the National Pain Foundation's division of therapies by degree of invasiveness:

Their continuum, "obeys a time-honored medical principle of using simpler, less invasive, and least costly interventions before using

Non-invasive Therapies	Invasive Therapies
Exercise	Medication management
Psychological pain management	Anesthetic blocking techniques
Physical and occupational therapies	Spinal cord stimulators
Biofeedback	Implanted pumps
Chiropractic manipulation	Peripheral nerve stimulators
Nutritional therapy	Surgery
Massage therapy	Chemical, surgical, or thermal nerve destruction
Psychotherapy	
Complementary medicine	

more invasive and more costly therapy. This plan suggests either using one therapy or more than one therapy at a time, abandoning those that do not work, and advancing to more invasive therapies, as in climbing a ladder." It starts with exercise, progresses through psychotherapy and over-the-counter medication, and, when all else fails, ends with highly invasive interventions such as "anesthetic nerve blocks, epidural steroid injections, implanted catheters, neurodestructive techniques, spinal cord stimulators, and morphine pumps."[16]

An example of a non-invasive, quick, and cheap intervention for anxiety is a self-administered ninety-second EFT routine. It takes under five minutes and involves absolutely no risk, yet it holds the potential to produce big shifts. If the problem does not shift after small-scale interventions, it might be necessary to scale up. That might mean enrolling in a class, taking a retreat, training in a specific technique, or trying several different energy therapies in quick succession to see if one of them brings relief. Finally, if all else fails, it might be necessary to scale up to an anxiety-reducing drug. But strong interventions like this should be the last resort, not the first. Hippocrates advised us to "First, do no harm," and that is a good first principle to start with in designing a "step-by-step plan" to address our healing issues.

4. Thrive Through Chaos

People often have elaborate sets of reasons as to why they cannot be well. Brad Blanton, Ph.D., a psychotherapist who wrote *Radical Honesty: How To Transform Your Life by Telling the Truth* and a number of other books, calls these our "tragic stories."[17] He explains how we get wrapped up in them and neglect the very things that are healthy and serving us in our lives right now.

Tragic stories prevent us from seeking our full potential, let alone realizing it. In *The Beethoven Factor,* Paul Pearsall, Ph.D., talks about how many great works of art, literature, and science have been produced despite the chaos in the personal lives of their creators.[18] Although he cites dozens of examples, he picks Beethoven as the archetypal creator of beauty amongst personal tragedy. Forsaking the traditional label of "survivors," he calls these people "thrivors," people who don't just make do, but go on to extraordinary accomplishment despite psychological, spiritual, and physical setbacks, tragic stories that any of them might have used as a valid excuse to give up hope and accept a limited life. In a condensation of *The Beethoven Factor* that Dr. Pearsall prepared for *The Heart of Healing,* in a section entitled "A Thrivors' Hall of Fame,"[19] he assembled the following partial list of thrivors, some of them world-famous, others obscure:

Lance Armstrong: He thrived through cancer to achieve unprecedented success in bike racing, and to inspire other cancer patients.

Poet William Carlos Williams: He suffered a severe stroke and subsequent emotional breakdown, only to later write great poetry and win the Pulitzer Prize for his work *Pictures from Brueghel.*

Nelson Mandela: He emerged from years of imprisonment and torture to become a leader for freedom, democracy, and the rights of the oppressed.

Pierre-Auguste Renoir: Unable to walk, and with fingers twisted by arthritis, he attached a paintbrush to his hand and painted some of the world's most memorable works, including (at age seventy-six) "The Washerwoman."

Henri Matisse: Suffering from heart failure, gastrointestinal disease, and with his lungs failing, he placed paintbrushes on a long stick and painted from his bed. His style created an entirely new field with a unique combination of color and form.

Enrico Dandolo: While serving as a peace ambassador to Constantinople in 1172 C.E., he was blinded in both eyes by the emperor's guards. Twenty-nine years later, and at age ninety-four, he led Venice to victory over Constantinople, and at age ninety-seven was appointed chief magistrate of Constantinople.

Sister Gertrud Morgan: She devoted her entire life to establishing and running an orphanage in New Orleans named Gentilly. When she was sixty-five years old, a hurricane destroyed her orphanage. She then returned to her interest in painting and went on to have her works displayed in museums around the world.

Ding Ling: (A pseudonym used by the Chinese novelist and radical feminist Kian Bingzhi.) She was imprisoned from the ages of sixty-six to seventy-one, during the Cultural Revolution of the 1970s in China. Upon her release, she went on to write some of her most highly praised works. She wrote an inspiring novel describing her experience of banishment to China's northern wilderness.

Helen Keller: Blind, deaf, and mute from age nineteen-months, she wrote and published (at age seventy-five) her book *Teacher* in honor of the woman who helped her thrive through her suffering.

Jesse J. Aaron: A descendant of slaves with a Seminole Indian grandmother, he too worked as a slave. Throughout his life he cared for his disabled wife and had to spend all of his meager funds on surgery to save his wife's sight. In poverty, he offered a definition of what I am calling the Beethoven Factor. He wrote, "It was then that the Spirit woke me up and said, 'Carve wood.'" He went on to become one of the most respected wood sculptors in the world.

The new model of wellness focuses on what's working in a patient's life, and what their potentials are, as well as treating the issues that trouble them. But the places of wellness will be the starting point, not the end point. I had a mentor, Bill Bahan, D.C., when I was taking my first classes in energy medicine in my late teens. One of his favorite sayings was, "What's right with you is the point. What's wrong with you is beside the point."[20]

Our lives are often messy, ambiguous, perplexing, and incomplete. Yet the doctor of the future will see, in every patient, in whatever state, the presence of the quantum potential of wholeness. That becomes the starting point for every treatment, for every step on the journey of recovery, regardless of whether the body and ego survive. Even if they do survive, it is just for a limited season, for eventually, *every* body and *every* ego will die. If our medical model is not wrapped up in attachment to how long that period is, and instead focuses on what's vibrant, alive, and vital in the person right now, it has a far more promising starting point for creating wellness. Once we let go of our tragic stories, the ground of healing is open to us. We can thrive in the midst of chaos.

5. Discover What Triggers Your Quantum Field

Years ago, I was a frequent visitor to a spiritual community in the Catskill Mountains of New York. It embraced a rigid lifestyle: meditations in the chapel before breakfast, study and classes in the morning, work in the afternoon, and worship at night.

There was another spiritual community close by, founded by Swami Muktananda. One day I visited there. The Swami was not in residence, but I talked to the chief administrator. "What techniques does the Swami advocate?" I asked, as we swapped notes about how our two communities were structured. "Oh, he might tell a student that he is forbidden to meditate at all," he replied. I was aghast. "Why would he say that?" I wondered.

"Perhaps the student has been meditating for hours each day for many years, and has become detached from the material world. But for another student, who has never meditated, he might have them sitting in lotus position for day and night."

The Swami was skilled at noticing peoples' habitual tendencies, and giving his students the opportunity to express their divinity in ways that were unfamiliar to them. Each of us needs to find our own answer to the question: "What activates the expression of my inner wisdom?"

The answer is absolutely unique for you, and you have to figure out what is best for your particular constellation of body, mind, and heart. It may look very different from the culture's vision—or even from your own beliefs.

For instance, in the early 1990s, many of my friends belonged to gyms and they sometimes tried to recruit me. But I lived twenty minutes' drive from the nearest gym. Driving there, dressing for exercise, working out, showering, and driving back, all collectively took about two hours. I just couldn't work that commitment into my day on a habitual basis. But what I could do was work out for twenty minutes a day at home. So I bought some home exercise equipment like a rebounder, a weight machine, an ab wheel, and some free weights, and used them regularly.

Working out at a gym fit well for my friends. My routine worked for me. Once you set the intention of being healthy, you have to discover what it is that works for you. Later, I moved to a house five minutes from a gym, joined, and began going regularly—and sold my home exercise equipment.

For me, forty-five minutes of meditation each morning before I work out is essential. It is like a drug, or rather an antidote to the slothful, addictive, and intellectually lazy tendencies I have. It took me years to realize this and give myself that time in the mornings.

I take a peculiar mix of supplements every day. They're the ones I notice work well with my physiology. I review the list and make changes every so often. My regimen might not work for you; yours might not work for me. Understanding our bodies and listening to their signals is a vital part of wellness.

What colors make you feel comfortable? What music soothes your heart? What images nourish your soul? What people do you have in your life that affirm the best in you? What events stimulate your creativity? Setting up our lives so that our highest potentials are continually affirmed by our external environments, and so that our outer world echoes the best of our inner world, is an invitation to the quantum universe to dance with us.

Eve Bruce, M.D., is a plastic surgeon who discovered indigenous healing during a trip to South America and was eventually initiated into the Yachak tribe as a shaman. She published a book about her experiences called *Shaman, M.D.* She beautifully articulates the call and response of the quantum universe: "In our culture we seem to have many answers. When asked why we had an accident or a disease, or in the face of global climate change, we give many answers—faulty tools, faulty user, genetics, biochemical and anatomic mishaps, pollution, the shrinking ozone layer. Yet these are answers to the question how, not why. 'Why' questions lead to a message. What is the message? What is spirit telling us through the language of our physical existence? How can we connect more fully to our physical existence and begin to hear God? The answers are within ourselves. We need only to ask, open up to the answers, and pay attention."[21]

Once you state your intentions, once you invite the quantum universe into a conversation about your wellbeing, listen intently for the answers. Find the mix that is just right for you. Don't be too swayed by the latest fads, but tune in to the whispers of your soul and the storehouse of advice it contains.

6. Allow for the X Factor

I used to write an operations plan for a $10 million book publishing and distribution company each year. It was a famous exercise within the company, involving every employee. The result was a 150-page operating manual that told everyone what everybody else was doing. But I also had a section called "The X Factor" with a couple of blank pages behind it. "You never know what's going to come up," I told curious and skeptical enquirers, "and while you can plan for every conceivable eventuality, there will always be occurrences you did not foresee." Quantum healing is a lot like that. You state your intentions, you fuel them with passion, and then you wait upon the universe to see what X factors it throws your way. Energy moves in mysterious ways, and you cannot graph the paths it might take.

Exceeding the Limits of Vision

Every human being is inclined, in the words of nineteenth-century philosopher Arthur Schopenhauer, to "take the limits of his own field of vision for the limits of the world." In 1899, Charles Duell, then U.S. Commissioner of Patents and Trademarks, urged the U.S. Congress to abolish his office, with the conviction that "Everything that can be invented, has been invented." His position seemed reasonable at the time: After all, for transportation, we had railroads that traversed continents, and steam ships that could cross the Atlantic in less than a week. For communication, we had telegraphs that could relay messages in Morse code around the world.

Yet within a half a century of his prediction, a flood of inventions—the airplane, the automobile, the telephone, the computer— had made nonsense of his prediction.

Today we are at a similar place in the history of healing. Modern technological medicine has given us great benefits, and will continue to produce breakthroughs. But it is dawning on us that the solutions we find outside of ourselves might be dwarfed in the magnitude of their effects by the solutions that lie dormant within our own consciousness.

For most of the last half of the twentieth century, scientists assumed that consciousness arose as the result of increasing complexity in living systems. They believed that as living systems evolved from bacteria, to simple animals, to complex animals, to mammals with large brains, they gradually evolved this phenomenon called "consciousness" to cope with the increasing complexity of life. This unexamined assumption pervades the writings of scientists. Stephen Jay Gould, one of the fathers of modern evolutionary biology, wrote that, "Humans arose, rather, as a fortuitous and contingent outcome of thousands of linked events, any one of which could have occurred differently and sent history on an alternative pathway that would not have led to consciousness."[22]

In the jargon of science, consciousness is regarded as an "epiphenomenon of matter." Matter comes first, then consciousness. In this conviction, consciousness is something that arises out of the interaction of trillions of molecules within your body. This view was encapsulated by Sir Francis Crick, a co-discoverer of the double helix structure of the DNA molecule, in his 1994 book *The Astonishing Hypothesis.*

He wrote that, "'You,' your joys and your sorrows, your memories and your ambitions, your sense of personal identity and free will, are in fact nothing more than the behavior of a vast assembly of nerve cells and their associated molecules. As Lewis Carroll's Alice might have phrased: 'You're nothing but a pack of neurons.'"[23]

This perspective certainly is astonishing, but for reasons other than the ones Sir Francis imagined. For science has taken us by the shoulders, turned us around, and pointed us resolutely in the direction opposite to his hypothesis. We now realize that consciousness underlies and organizes matter, and not the other way around. We have discovered that changes in consciousness precipitate changes in matter. We have realized that it is consciousness that is primary, and matter secondary.

The changes in our bodies produced by consciousness reveal the most potent tool for healing we have ever discovered. To a culture accustomed to looking for solutions "out there," it seems inconceivable that the answers might lie within. Yet scientific research, the very method we use to study the world "out there," now boomerangs back at us, pointing us to a much more challenging frontier. It reveals that the world "out there" is influenced by every shift we produce "in here." The stack of research piles higher every year. It shouts to us that the tools of our own consciousness—faith, prayer, optimism, belief, vision, charity, energy—hold promises of health, longevity, and peak performance that the interventions "out there" are hard-pressed to match.

An ancient Sufi story tells of the angels convening at the dawn of time to discuss where to bury the meaning of life, a secret so sacred that only the most worthy of initiates should be allowed access to it.

"We should put it at the bottom of the ocean," one exclaims.

"No, the highest mountain peak," argues another.

Eventually the wisest angel speaks up: "There is one place no-one will look. We can hide it in plain sight: in the center of the human heart."

Medicine, wellness, and healing look very different in a quantum world than they did in the mechanistic and reductionist world that preceded it. We see levels of causality and epigenetic control that supersede the code in our DNA. The leading edge of scientific research opens up a new creative realm, in which the intangible energy of our thoughts and emotions affects the tangible physical universe itself.

Each new discovery is another indication of how important consciousness is for healing. We are learning to see our cells and our bodies as malleable, influenced by every thought and feeling that flows through us. We can claim responsibility for the quality of thought and feeling we host, selecting those that radiate benevolence, goodwill, love, and kindness. Doing this, we are doing more than conscious epigenetic engineering on our bodies; we are nudging the whole world imperceptibly in the direction of vibrant health.

Appendix A:
How to Choose a Practitioner

If you have a condition that needs healing, and you've chosen soul medicine, you have a wealth of potential modalities and practitioners from whom to choose. You have access, through Ayurveda, Chinese medicine, and Shamanism, to therapies that have over twenty times the years of experience that modern Western medicine possesses. You have recourse to treatments like EFT and subconscious programming, which can produce very rapid change. You have options like electrostimulation and Healing Touch, which can improve conditions that baffle allopathic medicine.

Once you've researched soul medicine modalities, you are ready to choose a practitioner, or team of practitioners, to assist you on your healing journey. Making the right choices here can make an enormous difference. There may be only one local practitioner of the method you want to explore, or there may be dozens. It will require some work on your part to narrow your search. What is the best way to approach it?

First, faith in your practitioner is essential. It's even more important, given the nature of soul healing, than the modality you choose. Since we know that most disease results from blockages between soul and body, the method you use to reduce or remove the blockages is less important than the act of removing them. An acupuncturist can help create flow, as can an electromagnetic device. Since the quality of your own intention and belief is the most important single element in your healing, you must have a practitioner who you believe is able to help you effectively.

Second, having an active partnership with your practitioners is a must. You have responsibility for your own wellbeing, and you are seeking a partner who will assist you in your journey. Choose a practitioner in whom you trust, and with whom you feel comfortable. Communication should feel clear and easy. You should feel a mutual respect. It should be easy to be completely honest with this person. After all, you are forging a relationship between partners, a relationship on which your life may depend. A high-quality relationship creates an alliance to facilitate healing.

It's important to meet your practitioners in person before making final choices. You should feel comfortable in their professional setting. Compare your expectations with theirs. Check in with your body: How do you feel when you are around this person? Are there areas of tightness or strain? What are they trying to tell you?

Check out the background and qualifications of the practitioners you are considering. You will usually find one or more of the following qualifications:

Licensed Professionals

This group includes medical doctors (MDs), osteopaths (DOs), acupuncturists (L.Ac.s), and nurse practitioners. The licensing procedure for these professionals is rigorous. After several years of schooling, they are required to pass a stringent exam and complete a residency program under the supervision of highly experienced clinicians. However, the percentage of these professionals with an integrated, holistically oriented approach is small.

Certified Practitioners

Numerous bodyworkers and energy healers fall into this category. Examples of certification are Certified Massage Therapists (CMTs) and Certified Hypnotherapists (CHTs). Certification standards differ widely from state to state and profession to profession. A certification can mean as little as having completed an eight-hour class, or as much as having 3,000 hours of supervised practice. You can usually find out what's involved in obtaining certification in your state by doing a web search. This will acquaint you with the qualifications of the person from whom you are seeking treatment.

Healers

Healers may have studied a great deal, and worked under the supervision of teacher for many years. Or they may be self-described and self-ordained. They may have unique gifts that fit into no professional category. Most of the healers described in earlier chapters fit this description. They are bona fide spiritual healers whose gifts usually became apparent in childhood. Yet there are also charlatans and con artists in the business of taking advantage of sick people, so it's worth checking out an uncertified practitioner especially closely.

There are also practitioners who are knowledgeable in several fields. It is not uncommon today to find medical doctors who are also certified Ayurvedic physicians, licensed acupuncturists, and licensed naturopaths—all in one. Healers tend to have a lifelong interest in healing, and a curiosity about new approaches. Licensed professionals are also required to obtain continuing education credits, or CEs, each year. Many use this requirement to study an additional modality.

Master Zhi Gang Sha was already a highly respected physician and acupuncturist when he apprenticed himself to Master Guo. He writes: "I earned a medical degree in Western medicine at Xian Jiao Tong Medical University. However, my work as an institutional physician soon revealed that Western medicine was unable to help many patients. I realized that integrating it with traditional Chinese medicine would combine the best of both systems, so I obtained certification as a doctor of traditional Chinese medicine. [Then] I became a disciple of Dr. Zhi Chen Guo. Under Master Guo's rigorous training...my spiritual channels were opened, and I became a medical intuitive."[1]

A healer may have studied a particular modality but not have the experience to treat you in depth; in such cases ethical practitioners recognize the limits of their expertise and will certainly refer you to a highly competent specialist. Throughout this book, and in the appendices, you will find the names of healers we recommend that you investigate.

Referrals and Office Visits

Once you've checked out the qualifications of a potential healing partner, ask for referrals. Most healing professionals will supply you with the names of past patients who can vouch for their work. Your coworkers, family members, friends, and social contacts are also sources of possible information. Your primary care physician may have suggestions for soul medicine practitioners, and may surprise you with his or her knowledge of the field. Patients groups, support groups, and professional societies can supply you with information. Most colleges have lists of their practicing alumni online. Groups like the American Holistic Medical Association (www.AHMA.com) and the American Board of Scientific Medical Intuition (www.ABSMI. com) make their practitioner lists available online, and you will find many others in almost every specialty with the aid of a web search engine.

Practitioner web sites are also a good source of information. Their web sites will often tell you their credentials, professional affiliations, training, how many years they've been practicing, treatment philosophy, endorsements and testimonials, details of their particular techniques, and what you can expect during the course of a professional relationship.

Use a search engine and type in the practitioner's name. This will give you a list of their publications, recent news stories about them, web sites on which they're listed, and a host of details. Patient complaints will often show up here; there are sites that allow people to let off steam about bad experiences.

Payment information is usually not on a practitioner's web site, and you will usually have to call their office to find that out. A few operate within HMO, PPO, or other medical insurance plans; most do not. Many offer a sliding scale based on financial ability. Some offer a free or low-cost introductory visit or examination.

Use your intuition. Remember that your 11+ million bit-per-second subconscious scanning computer is already doing the evaluation! Draw on its wisdom. An initial consultation, armed with the right questions, will give you a sense of whether this is the right healing partner for you.

In the course of that initial consultation, be completely frank with your prospective partner. Describe all aspects of your life that might be relevant to your health concern. Make your expectations explicit. Ask the practitioner for a frank assessment of your future prospects with them. If anything the person says doesn't sound right, let them know, and take note of how they handle the interaction. Are they able to accept feedback? Notice the person's experience and intelligence level as well as the feel-good factors; there are highly empathetic practitioners who are not very effective, and geniuses are often short of manner, and intolerant of lesser talents.

Find out whether they will keep written records, and whether they will share them with other healing partners. Find out how long they estimate your treatment might take. Make notes, after the interview, of your observations, both positive and negative. If you don't already keep a journal, a healing journey is a great opportunity to start one. Practitioner interviews will focus you on tuning into your body and intuitive wisdom, as well as your intellectual observations.

Get an accurate diagnosis. This might take visits to several different practitioners. But the ability to diagnose illness is one of the crowning achievements of Western medicine, and the time and expense required is virtually always worthwhile. An accurate diagnosis is an essential first step in treatment. For some conditions, especially those related to EMD, a conventional allopathic diagnosis might be difficult to obtain. But a visit to a good diagnostician can certainly rule out obvious organic causes of disease.

Supplementing a *diagnosis* by a licensed professional is the possibility of an *assessment*. An assessment is done by a healer; rather than aiming to identify a disease, it targets an area of energy blockage. The site of the blockage may be far from the site of symptoms, and precede any symptoms at all. Daniel Benor, M.D., a distinguished researcher who was the first to systematically catalog scientific studies that supported the role of spirituality in healing, drew this distinction in a paper entitled *Intuitive Assessments: An Overview*. He writes: "Reports abound of healers identifying problems present in healees. Many healers are able to intuit where to place their hands in order to give healing. Healees often comment on the fact that healers "find the right spot" without being told.... Some healers feel or see the biological energy fields around people and are guided by their sense of touch or by the colors of the energy field to places that need healing.... I have spoken with hundreds of healers and healees over two decades of healing explorations. I cannot count the numbers of stories I have heard of chronic pains, fevers, and diseases that eluded medical diagnosis and were clarified through intuitive assessments."[2]

Once you start treatment, remember the partnership model. You and the practitioner or practitioners are a team, committed to your healing. Be proactive. Ask questions. Challenge suggestions that you don't understand, or that feel wrong to you. Share disappointments as well as triumphs. Talk about your expectations openly. Get your practitioner's email address so that you can communicate in this way too, and find out how often he or she checks and responds to emails. But don't impose, expecting your practitioner to substitute long free email conversations for office visits. They do not have the time.

Let all the members of your healing team know what's happening. You might set up an online group to share information; several portals like Yahoo, Google and MSN have "group" services that take just a few minutes to set up. Having a group like "GeorgeGoringHealing@yahoogroups.com"—and entering the email addresses of all those on your healing team, allows a comment sent by any practitioner to go to all. Again, take care about email contact. It's like a hot line: used sparingly, it can be vital. If you overuse it, your healing professionals will tend to ignore your communications in this format.

Keep your primary care physician informed about other treatments. Patients often make the mistake of thinking that herbs must be safe because they're natural. But some herbs have adverse reactions when used with certain drugs. And drugs can inhibit the effectiveness of some herbal formulations. Your N.D., who may be prescribing herbs, needs to know what your M.D., who may be prescribing drugs, is doing, and vice versa. An informed treatment team is an effective treatment team.

Sometimes even the smartest patient simply cannot understand some aspect of treatment. You might feel foolish asking, or asking again—and again.

Do it anyway. Healing professionals have their own mindsets and jargon. They may think they're being perfectly clear, and the course of treatment might indeed be routine and simple to them. But it's new to you, and it might take you a while to understand aspects of it. So nodding your head when you can't follow the conversation is a bad idea. Ask until you're sure you understand what's going on—it's an essential part of taking responsibility for your own wellbeing. You are your best physician. One of the most wonderful integrative, holistic physicians, Gladys McGarey, M.D., often speaks of the physician within. If you realize that your problems are primarily the result of unresolved emotional stress, you will find a complete self-therapy program in *Ninety Days to Stress-Free Living,* a guide to nurturing yourself.[3]

Use your journal to track the results of your treatment. Evaluate how you're doing. Have the patience to give each treatment time to have its effect. Some produce very quick results. Some may result in you feeling worse before you feel better. Getting in touch with your body, for example—and really being present to the dysfunction that you've been pushing to the back of your mind for the last five years—may feel very uncomfortable for a while. Holistic health aims to treat the whole person, rather than aiming a magic bullet at a particular problem. Your practitioner may treat a site a long way from the pain. Soul medicine looks for the root problem, rather than trying to make the symptom go away. Your practitioner can usually give you an average time it has historically taken other patients with your condition to get well—your healing time may be more or less. It may take patience, and learning patience is an essential part of a healing journey.

The Old Testament tells the story of Naaman, head of the mighty Syrian army. He developed leprosy, and his body was covered with sores. His healing instruction was to, "Go and wash in Jordan [river] seven times, and thy flesh shall come again to thee, and thou shalt be clean." Naaman, an impatient overachiever, "was wroth, and...went away in a rage." He had many better ideas of how his healing might be accomplished. Yet when he finally learned patience, and went along with the program, "his flesh came again like unto the flesh of a little child,"⁴ and he was healed. So give the process time, and keep an open mind.

A healing partnership involves both information and intuition. Illness is an opportunity to become exquisitely conscious, to straddle both worlds, to be fully in your body as well as fully present with your spirit. It is a chance to find out about advanced treatments as well as using common sense. Healing is a milestone on your spiritual journey. Embrace it as a powerful spiritual teacher.

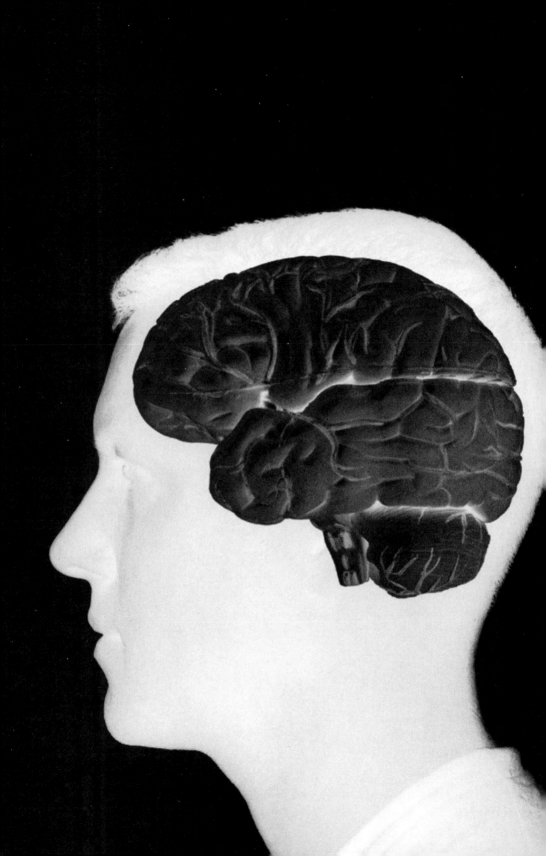

Appendix B: EFT Basic Recipe

Preliminaries: Balance energies, select problems, rate problem from 1 to 10, word Reminder Phrase.

Part 1—Setup: Rub chest sore spots or tap karate-chop points while saying three times, "Even though [name problem], I deeply love and accept myself."

Part2—Tapping: Tap the points described below while saying your Reminder Phrase out loud.

Part 3—Nine-Gamut Procedure: Tap the point between the little and fourth fingers, wrist side of the knuckle, as you: 1) close your eyes, 2) open your eyes, 3) look down to the right, 4) look down to the left, 5) circle your eyes, 6) circle your eyes in the opposite direction, 7) hum a bar of a song, 8) count to five, 9) hum again. Optionally, end by sweeping your eyes out and up, sending energy through them.

Part 4—Tapping: Repeat Part 2.

Repeat this sequence until your rating of the problem is a 0 or near 0. Challenge the results by attempting to invoke the disturbing feeling. Once you cannot create the unwanted emotional response, you are ready to test the gains in a "real life" setting.

If the Problem Is Not Responding, identify and address 1) other aspects of the problem, 2) psychological reversals, 3) scrambled energies, or 4) energy toxins.

The Tapping Points:
Beginning of the Eyebrows
Sides of the Eyes
Under the Eyes
Under the Nose
Under the Lower Lip
K-27 Points
Arm-attachment Points
   (optional)
Thymus Thump (optional)
Under the Arms
Outside of the Legs (optional)
Karate-chop Points

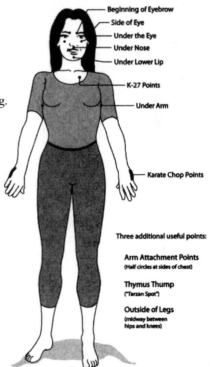

EFT Tapping Points

327

Appendix C:
Five Minute Energy Routine

The Five Minute Energy Routine packs many of the principles of energy medicine and Energy Psychology into a compact, practical form. It can be used in the morning to center yourself before starting your day, or when you're feeling stressed. This set of techniques also works well any time you are feeling tired or scrambled, can't think clearly, feel hysterical or out of control, feel droopy and out of energy, or need a pick-me-up. It represents the distilled wisdom of a quarter century of practice by the gifted energy healer Donna Eden, and is presented with more discussion in her book *Energy Medicine*. This abbreviated version is used with her permission; for a better understanding of the derivation of the techniques, as well as a host of other methods for use in specific situations, *Energy Medicine* is a valuable guide.

The Five Minute Energy Routine consists of seven postures and movements, and takes under five minutes to perform. Note how you feel in your body before doing the routine, and how you feel afterwards. Most people who practice the routine feel a marked shift in the five little minutes required to complete the process. The routine includes:

1. The Three Thumps
2. The Cross Crawl
3. The Wayne Cook Posture
4. The Crown Pull
5. Neurolymphatic Massage
6. The Zip Up
7. The Hook Up

Here's how you do the routine:

1. The Three Thumps

Tap or thump the indicated collarbone points with the tips of your four fingers, or the front of your fist, for about twenty seconds. Then tap over your thymus gland.

Collarbone points

Thymus point

2. The Cross Crawl

The Cross Crawl looks like marching in place. You swing your left arm up whilst raising your right leg, and vice versa. Move your arms in an exaggerated arc up and down, whilst also crossing the midline of your body.

Cross Crawl

3. The Wayne Cook Posture

Sit down and place your right leg over your left knee. Wrap your left palm around your right ankle. Wrap your right palm around your instep. Breathe slowly in through your nose, and out through your mouth for the duration of this pose.

Do the mirror image of this posture.

Place your thumbs on the bridge of your nose, with your fingertips together, and take three deep breaths.

Right foot hold

Left foot hold

Center hold

4. The Crown Pull

Place your fingertips on the centerline of your forehead. Using moderate pressure, pull them apart. Repeat this pull several times, each time moving your fingers up, till you've moved over the top of your head, and are pulling at the back of your skull.

Crown pull

333

5. Neurolymphatic Massage

Massage the following places with your fingertips:

Under your collarbone

Where the fronts of your arms connect to your torso

Down the front of your breastbone

Under your breasts

Under your ribcage, an inch to each side of your breastbone

The outside edge of each thigh, from the hips to the knees

The back of your neck, from the base of your skull, then traveling as far down as you can reach.

6. The Zip Up

Place your fingertips on your pubic bone.

Take a deep inbreath, whilst moving your hand up the center of your body, to your lower lip.

Repeat two more times.

7. The Hook Up

Place a fingertip in your navel.

Place a fingertip of your other hand between your eyebrows.

Pull up gently with both fingertips as you take a deep breath.

That's it! Five minutes have gone by, and most people find themselves rejuvenated. After you've tried out the Five Minute Energy Routine once or twice, you'll find it useful anytime you need a quick boost, to center yourself at the beginning of your day's activities, or when you're about to enter a situation you suspect will be stressful. And beyond the immediate benefits, the routine, when practiced regularly, is designed to establish healthier "energy habits" throughout your body.

Appendix D:
About Soul Medicine Institute

S oul Medicine Institute is a nonprofit organization dedicated to encouraging practices and attitudes that facilitate the under-standing that true wellness starts with a vibrant spiritual con-nection. The manifesto of Soul Medicine is contained in the book of that title by Norman Shealy, M.D., Ph.D., and Dawson Church, Ph.D. The book *Soul Medicine* is a comprehensive guide to the history and principles of the emerging disciplines of energy medicine and sacred healing. It summarizes over one hundred scientific studies, and dozens of medically proven miraculous cures, that demonstrate the power of Soul Medicine. It identifies three pillars of Soul Medicine. They are:

Energy
Intention, and
Consciousness

It identifies therapies that work at the level of the body's electro-magnetic energy system. It also shows how intention, projected into the energy system by consciousness, can produce miracles of healing. The research behind Soul Medicine illuminates links between gene expression, electromagnetic energy transfer and signaling, quantum mechanics, string theory, and human consciousness. It demonstrates a sound theoretical framework, based on credible experiments, for understanding these healing breakthroughs, and predicts that these techniques will dramatically advance the fields of medicine and psy-chology in the coming decade.

Therapies and Practitioners

Through the web site at www.SoulMedicineInstitute.org, you can find links to the web sites of each of these new therapies. You can also find lists of practitioners certified in these methods.

Educational Institutions

This body of knowledge is being systematically taught at many educational institutions. There are now degrees available, including advanced degrees, at several of them. Some are accredited, while others are not. A list of some of the most comprehensive programs is found on the Soul Medicine Institute site.

Scientific Research

Besides distributing such information, Soul Medicine Institute is dedicated to furthering our understanding of these techniques through sound scientific research. It is currently involved in several studies of the effects of Energy Psychology interventions, and enthusiastically welcomes volunteer support and funding for these efforts. Contact information for volunteer opportunities and financial contributions can be found on the web site.

Trainings Offered

Soul Medicine Institute also offers training in these methods for groups, organizations, and companies. These techniques have been shown to alleviate stress much more quickly than other psychotherapeutic approaches. The most popular training is called the Three Minute Stress Dump (www.StressDump.com). In addition to explaining the science behind Energy Psychology, it trains participants in methods of alleviating stress in just a few minutes. These practical techniques have been used successfully to:

Alleviate workplace stress
Decrease workplace injuries
Reduce anxiety in particularly stressful situations

Eliminate phobias such as fear of public speaking, or heights
Reduce PTSD (Post-Traumatic Stress Disorder)
Improve marital relationships
Eliminate allergies
Alleviate depression
Produce remission in serious diseases such as cancer

Case Histories

Soul Medicine Institute has also set up the first international database of Energy Psychology case histories. This research tool collects medical and psychiatric diagoses before treatment. It notes the Energy Psychology treatments used, and the diagnosis after treatment. It is peer reviewed, and conforms to the Consort Standards and the standards of the National Institutes of Health.

The database is very user-friendly, and accessible on the web at any time. Patients, doctors, and other clinicians may submit cases for review. If you have been treated using an Energy Psychology method, or if you have a healing story to share, consider writing it up for consideration in the database, or encourage your doctor to do so. A direct link to the database can be found at www.CaseHistoryDatabase.org.

Grants, Bequests, and Funding

Scientific research requires money! Because it does not benefit a drug company, hospital, or much of the infrastructure of the current medical system, so far there has been only a tiny trickle of funding for Energy Psychology and energy medicine research.

Soul Medicine Institute has a number of studies underway or planned that need funding. Please consider making tax-deductible a donation or bequest.

For more information, go to www.SoulMedicineInstitute.org

Appendix E:
Research at Soul Medicine
Institute

S oul Medicine Institute is involved in a number of exciting research initiatives. Among these are:

MACE Study

The MACE Study (Mitigation of Adverse Childhood Experiences) examines the effects of Energy Psychology treatments on people suffering the effects of adverse childhood experiences. A ten-year study by Kaiser Permanente of over 14,000 adults (the ACE study) found that there is a high degree of correlation between adverse childhood experiences and adult disease, including heart disease, cancer, hypertension, diabetes, and depression.

The MACE Study examines the health effects of the reduction in adult stress that comes from using Energy Psychology treatments. It hypothesizes that if adult disease correlates highly with childhood stress, then the amelioration of stress might bring about a corresponding decline in disease rates. The MACE Study has been proposed by Soul Medicine Institute as a collaboration with Kaiser Permanente, the Association for Comprehensive Energy Psychology, and other institutions.

Clinical Case History Database

Thousands of doctors, psychotherapists, psychiatrists, social workers, and other mental health professionals use EFT and other Energy Psychology methods in their practices. And hundreds of thousands of people subscribe to the EFT web site. Every month, hundreds of them post success stories on the web.

The Clinical Case History Database (CHD) is a research project funded, sponsored and maintained Soul Medicine Institute. The CHD examines accounts of healing or remission that result from Energy Psychology treatments. It rigorously evaluates them for medical accuracy, according to standards established by the National Institutes of Health (NIH), the Ethics Guidelines of the Association for Comprehensive Energy Psychology (ACEP), and other professional bodies. Those that pass the screening and peer review process are posted on the web, after personal identifying information has been removed. The information in the CHD is freely available to researchers and the public without charge.

Anyone can post a proposed case history on the CHD. If you have had an experience with Energy Psychology, or your provider has used it in your treatment, you are encouraged to log onto the web site and write up these experiences, or encourage your provider to do so. The direct link to the CHD can be found at www. CaseHistoryDatabase.org.

PTSD Study

Soul Medicine Institute has funded a proposed study of civilian contractors recently returned from Iraq, to determine the effects of Energy Psychology on Post-Traumatic Stress Disorder (PTSD).

Phobia Study

Soul Medicine Institute is working with other groups to initiate a study at Kaiser Permanente to determine the effects of Energy Psychology on phobias. This is an extension and replication of the two phobia studies described in Chapter 11, with potential for triggering a larger-scale study in the Kaiser system.

Cortisol/Catecholamine Reuptake Study

Cortisol and catecholamines are "stress chemicals" measurable in blood samples. Concentrations of cortisol and catecholamines increase in people under stress, and decrease as stress decreases. Once a stress is removed, these molecules are broken up by the body. Catecholamines are disassembled in a few minutes by enzymes, especially monoamine oxidase.

Further, the same precursor hormone is used by the body to manufacture both cortisol and DHEA, the most common hormone in the body, and one that is known to have anti-aging effects. When the production of cortisol is interrupted and cortisol levels drop, production of DHEA rises proportionately. The reduction of cortisol and the corresponding increase in DHEA production has regenerative effects on many of the body's systems, countering the effects of aging.

The Cortisol/Catecholamine Reuptake Study has been proposed by Soul Medicine Institute to study whether these stress chemicals are broken down more rapidly by the body after an Energy Psychology treatment. If Energy Psychology results in a greater reduction in stress chemicals than other treatments for stress, it will demonstrate an important physiological effect of these treatments. It will also imply increased expression of the genes involved in coding for the proteins involved, and a link between the piezoelectric effects of tapping and gene expression.

Gene Chip Studies

Soul Medicine Institute also proposes studies that examine the expression of DNA sequences linked to the stress response, using gene chips. These chip studies would note which genes were expressed before Energy Psychology treatments, and which were expressed after.

ACEP Research Committee

The Research Committed of the Association for Comprehensive Energy Psychology (ACEP) encourages studies, case histories, and other initiatives that further research into Energy Psychology. More information can be found at the ACEP web site, www.energy psych.org.

Peer Review

Soul Medicine Institute plays a role in reviewing scientific papers prior to publication. The process of reviewing proposed research for design flaws, and papers for scientific accuracy, is an important step in making sure that high-quality data is generated and published.

Updates

Any information printed in a book becomes obsolete before it is off the press! Regular updates on all of these initiatives may be found at the Soul Medicine Institute web site, www.SoulMedicine Institute.org.

Funding

Compared to the flood of research money available for allopathic treatments, the funding for the safe and noninvasive therapies described in this book is the tiniest of trickles. Funds are urgently needed to carry forward the research above. If you or anyone you know has been helped by these treatments, consider making a tax-deductible donation to support research at Soul Medicine Institute, the ACEP Research Committee, or any other body investing in research into these therapies of the future.

Endnotes

CH. 1 EPIGENETIC HEALING

1. Rossi, Ernest (2002). *The Psychobiology of Gene Expression* (New York: Norton), p. xvii.
2. Ibid, Rossi, p. 9.
3. Kandel, Eric (1998). A new intellectual framework for psychiatry? *American Journal of Psychiatry,* 155, p. 457.
4. Kemperman, G., Gage, F. (1999) New nerve cells for the adult brain. *Scientific American,* 280, p. 48.
5. Greenfield, Susan (2000). *The Private Life of the Brain* (New York: John Wiley), 9.
6. Eldridge, Niles (2004). *Why We Do It* (New York: Norton), p. 15.
7. *HSC Weekly,* (1998, July 10). University of Southern California Health Science Campus, p. 3.
8. Nelkin, Dorothy (1995). *The DNA Mystique* (New York: Freeman), p. 193.
9. Justice, Blair (2000). *Who Gets Sick?* (Houston: Peak), p. 63.
10. Idler, E., & Kasl, S. (1991). Health perceptions and survival: do global evaluations of health really predict mortality? *Journal of Gerontology,* 46 (2), p. S55.
11. Dossey, Larry (2006). *The Extraordinary Healing Power of Ordinary Things.* (New York: Harmony), p. 21.
12. Wright, William (1998). *Born That Way: Genes, Behavior, Personality* (New York: Knopf), p. 13.
13. Dugatkin, Lee (2000). *The Imitation Factor: Evolution Beyond the Gene* (New York: Free Press), p. 8.
14. Ratner, Carl (2004). Genes and psychology in the news, *New Ideas in Psychology,* C.
15. Wilson, E. O. (2000). *Sociobiology: The New Synthesis,* Twenty-Fifth Anniversary Edition (Cambridge: Harvard University Press), p. 575.
16. Ibid, Eldridge, p. 22.
17. Crick, Francis (1970, Aug 8). Central dogma of molecular biology. *Nature,* 227, p. 561.
18. Ho, Mae-Wan (2004, March 9). Death of the central dogma, Institute of Science in Society press release.
19. Fisher, Charles (2007). *Dismantling Discontent: Buddha's Path Through Darwin's World* (Santa Rosa, Elite Books), p. 177.
20. McCallum, Ian (2005). *Ecological Intelligence,* (Cape Town: Africa Geographic), p. 5.
21. Oschman, James (2006). Trauma energetics. *Journal of Bodywork and Movement Therapies,* vol. 10, p. 21.
22. Rossi, Ernest (2002). *The Psychobiology of Gene Expression* (New York: Norton), p. 66.
23. Cousins, Norman (1989). Beliefs become biology. *Advances in Mind-Body Medicine,* 6, p. 20.
24. Richardson, K. (2000). *The Making of Intelligence* (New York: Columbia University Press), summarized in Rossi, p. 50.
25. Grauds, Constance (2005). *The Energy Prescription* (New York: Bantam).
26. Cool, L. C. (2004, May). The power of forgiving. *Reader's Digest,* p. 92.
27. Childre, L., & Martin, H. (2000) The Heartmath Solution (San Francisco: HarperSanFrancisco), p. 57.
28. *USA Today* (2005, March 8). Study: happy marriage heals.
29. Cahill, L. et. al. (1994). B-andregenic adtivation and memory for emotional events. *Nature,* 271 (2), p. 702.
30. Ibid, Grauds, p. 38.
31. Ibid, Rossi, p. 229.
32. Pennisi, E. (2001). Behind the scenes of gene expression. *Science,* 293, p. 1064.
33. *Science* (2001, August 10), *Epigenetics* special issue, 293, p. 5532.
34. Dossey, Larry (2001). *Healing Beyond the Body.* (Boston: Shambhala), p. 235.
35. Gaugelin, Michael (1974). *The Cosmic Clocks* (New York: Avon).
36. Shealy, C.N., and Church, D. (2006), *Soul Medicine* (Santa Rosa: Elite), p.177
37. Snyderman, Ralph (2005, October-December). Creating a culture of health. *Spirituality and Health Magazine,* p. 46.
38. For an extended discussion of these figures and their sources, see Shealy & Church, chapter 17.
39. Dworkin, Ronald (2005), Science, faith and alternative medicine. *Policy Review,* Issue 108.
40. Shealy & Church, p. 234.

41. Bader, C., et. al. (2006, September). American Piety in the 21st Century. *Baylor Religion Survey,* p. 45

42. Astin, J., et. al. (2003). Mind-body medicine: state of the science, implications for practice. *Journal of the American Board of Family Practice,* 16, p. 131.

43. Ibid, Shealy & Church, p. 235.

44. Folkman, Judah. (2001, February 19). From Kalb, C., Folkman looks ahead. *Newsweek,* p. 44.

CH. 2 YOU: THE ULTIMATE EPIGENETIC ENGINEER

1. Rabinoff, Michael (2007). *Ending the Tobacco Holocaust.* (Santa Rosa: Elite), p. 212.

2. Marcus, G (2004). Making the mind: Why we've misunderstood the nature-nuture debate. *Boston Review,* (Dec/Jan).

3. Carlson, Richard, in Kelly, Eamon (ed) (2002). What's Next (Cambridge: Perseus), p. 204.

4. Bultman, S. J., et. al., (1992). Molecular characterization of the mouse agouti locus, *Cell,* Vol. 71, p. 1195.

5. Watters, Ethan (2006, Nov). DNA is not destiny: the new science of epigenetics rewrites the rules of disease, heredity, and identity. Vol. 27, No. 11.

6. Weinhold, R. (2006, March). Epigenetics: the science of change. Environmental Health Perspectives, vol. 114 no. 3

7. *The Economist* (2006, September 23). Learning without learning. p. 89.

8. Steinberg, Douglas (2006, Oct/Nov), Determining nature vs. nurture: molecular evidence is finally emerging to inform the long-standing debate. *Scientific American Mind*, p. 12.

9. Marcus, G (2004). Making the mind: Why we've misunderstood the nature-nuture debate. Boston Review, (Dec/Jan).

10. Felliti, V. J., et. al. (1998, May 14). Relationship of childhood abuse and household dysfunction to many of the leading causes of death in adults. The Adverse Childhood Experiences (ACE) study. *American Journal of Preventive Medicine,* 4 p. 245.

11. Ironson, G., Stuetzle R., Fletcher M. A., and Ironson D. View of God is Associated with Disease Progression in HIV. (2006). Paper presented at the annual meeting of the Society of Behavioral Medicine, March 22–25, 2006 San Francisco. Abstract published in *Annals of Behavioral Medicine,* 31 (Supplement), S074.

12. Ironson G., Kremer H., and Ironson D. (2007). Spirituality, spiritual experiences, and spiritual transformations in the face of HIV. In: Koss-Chiono J., and Hefner, P. (eds.) *Spiritual Transformation and Healing* (Walnut Creek: Altamira Press)

13. Ibid. Ironson, G., et. al.

14. Donnelly, Matt (2006, May). Faith boosts cognitive management of HIV and cancer. *Science & Theology News,* p. 16.

15. Ironson, G., et. al. (2002). The Ironson-Woods Spirituality/Religiousness Index is Associated with long survival, health behaviors, less distress and low cortisol in people with HIV/AIDS. *Annals of Behavioral Medicine.* 24(1), p. 34.

16. Ironson G., Stuetzle R., and Fletcher M.A. (2006). An increase in religiousness/spirituality occurs after HIV diagnosis and predicts slower disease progression over 4 years in people with HIV. *Journal of General Internal Medicine,* 21, S62-S68.

17. Interview with author, Jan 30, 2007.

18. Grauds, Constance (2005). *The Energy Prescription* (New York: Bantam), p. 45, quoting *Journal of Gerontology,* 55A, p. M400.

19. Bader, C., et. al. (2006, September) *American Piety in the 21st Century.* Baylor Religion Survey. p. 26.

20. Ironson G., Kremer H., and Ironson D. (2007). Spirituality, spiritual experiences, and spiritual transformations in the face of HIV. In: Koss-Chiono J., and Hefner, P. (eds.) *Spiritual Transformation and Healing* (Walnut Creek: Altamira Press).

21. Bower, Bruce (2007, January 27). Mind over muscle: placebo boosts health benefits of exercise. *Science News Online,* 171 (4).

22. Oxman, Thomas E., et al. (1995). Lack of social participation or religious strength and comfort as risk factors for death after cardiac surgery in the elderly. *Psychosomatic Medicine,* 57, p. 5.

23. Powell, L. H., Shahabi, L., Thoresen, C. E. (2003, January). Religion and spirituality: linkages to physical health. *American Psychologist,* 58 (1), p 36.

24. Astin, J. E., et. al. (2000). The efficacy of "distant healing": a systematic review of randomized trials. *Annals of Internal Medicine,* 132, p.903.

25. Jonas, W. B. (2001). The middle way: Realistic randomized controlled trials for the evaluation of spiritual healing. *The Journal of Alternative and Complementary Medicine,* 7 (1), p 5-7.

26. Casatelli, Christine (2006, May). Study casts doubt on medicinal use of prayer: cardiac patients show no improvement after intercession. *Science & Theology News,* p. 10.

27. McTaggart, Lynne (2002). *The Field.* (New York: Quill), p. 189.

28. Dossey, Larry (1997). *Prayer is Good Medicine* (New York: HarperCollins), p. 104.

29. Wallis, Claudia (2005). The New Science of Happiness, *Time,* Jan 15.

30. Grauds, Constance (2005). *The Energy Prescription.* (New York: Bantam), p. 136.

31. Ibid, Donnelly, p. 16.

32. For a database of meditation studies, see the Institute of Noetic Sciences web site, specifically http://www.noetic.org/research/medbiblio/index.htm (accessed January 3, 2007).

33. Dozor, Robert. (2004). Integrative health clinic meets real world: *The Heart of Healing,* D. Church (ed). (Santa Rosa: Elite Books), p. 312.

34. Weil, Andrew (2006, July). *Dr. Andrew Weil's Self-Healing,* p. 2.

35. Ellison, Katherine (2006, September - October). Secrets of the Buddha mind. *Psychology Today,* p. 74.

36. Hirschberg, C. (2005) Living With Cancer, in *Consciousness & Healing* (St. Louis: Elsevier), p. 163.

37. Rhodes, D., et. al. (2007, June). Large-scale meta-analysis of cancer microarray data identifies common transcriptional profiles of neoplastic transformation and progression. *Proceedings of the National Academy of Sciences of the United States of America.* Published online before print. 101 (25), p. 9309.

38. Trudeau, Michelle (2007, Feb 15). Students' view of intelligence can help grades. NPR Your Health.

39. Weil, Andrew (2006, September). Attitude is everything with aging. *Andrew Weil's Self Healing Newsletter,* p. 1.

40. Shealy, C. N. (2005). *Life Beyond 100* (New York: Tarcher), p. 37.

41. Chidre, L., & Martin, H. (2000). The HeartMath Solution. (San Francisco: HarperSanFrancisco), p. 90.

42. Ibid, Childre & Martin, p. 55.

CH. 3 THE MALLEABLE GENOME

1. Silvia Hartmann was the therapist.

2. Feinstein, D., Craig, G., Eden, D. (2005). *The Promise of Energy Psychology* (New York: Tarcher), p. 109.

3. Rossi, Ernest (2002). *The Psychobiology of Gene Expression* (New York: Norton), p. 36.

4. Adapted from Wen, X. et. al. (1998). Large-scale temporal gene expression mapping of central nervous system development. *Proceedings of the National Academy of Sciences,* 95, p. 334.

5. Chen, Eachou (2006, November). Detection of differentially expressed genes using DNA microarray in acute tremor by electro-acupuncture treatment: the role of meridian system to conduct electricity and to transmit bioinformation. *Journal of Accord Integrative Medicine,* Vol. 2, no. 6, p. 156.

6. Bentivoglio, M., et. al. (1999). Immediate early gene expresion in sleep and wakefulness, in *Handbook of Behavioral State Control.* Lydic, Ralph and Baghdoyan, Helen A., eds. (New York: CRC), p. 235.

7. Tolle, T.R., et. al. (1995). *Immediate Early Genes in the CNS* (New York: Springer-Verlag).

8. Marcus, G (2004). Making the mind: Why we've misunderstood the nature-nuture debate. *Boston Review,* (Dec/Jan).

9. Glaser, R., et. al. (1990). Psychological stress-induced modulation of interleukin 2 receptor gene expression and interleukin 2 production in peripheral blood leucocytes. *Archives of General Psychiatry,* 47, p. 707.

10. Glaser, R., et. al. (1993). Stress-associated modulation of proto-oncogene expression in human peripheral blood leukocytes. *Behavioral Neuroscience,* 107, p. 525.

11. Castes, M., et. al. (1999). Immunological changes associated with clinical improvement of asthmatic children subjected to psychosocial intervention. *Brain and Behavioral Immunology,* 13 (1), p. 1.

12. Morimoto & Jacob, quoted in Rossi, p. 236

13. Ibid, Rossi, p. 236.

14. Ibid, Rossi, p. 237. Diagram used with permission.

15. Ibid, Rossi, p. 68. Diagram used with permission.

16. Ibid, Rossi, p. 57.

17. Werntz, D., et. al. (1987). Selective hemispheric stimulation by unilateral forced nostril breathing. *Human Neurobiology,* 6, p.165.

18. Ibid, Rossi, p. 68.

19. Ibid, Rossi, p. 68.

20. Ibid, Rossi, p. 103.

21. Ibid, Rossi, p. 103.

22. Ibid, Rossi, p. 202.

23. Kemperman, G., Gage, F. (1999). New nerve cells for the adult brain. *Scientific American,* 280, p. 48.

24. Ibid, Rossi, p. 229.

25. Ibid, Rossi, p. 191.

26. Weil, Andrew (2006, July). *Dr. Andrew Weil's Self-Healing,* p. 2.

27. Damasio, R., et. al. (2000). Subcortical and cortical brain activity during the feeling of self-generated emotions. *Nature Neuroscience,* 3, p. 1049.

28. Raheem, Aminah (1991). *Soul Return* (Connecticut: Aslan), p. 81.

29. Moyer, C.A., et. al. (2004). A meta-analysis of massage therapy research. *Psychological Bulletin,* 130 (1) p. 3.

30. *Washington Post* (2006, May 22). Botox lifts brows—and spirits.

31. Ibid, Rossi, p. 304.

32. Rossi, Ernest (2002). *The Psychbiology of Gene Expression,* p. 47.

33. Kandel, Eric (1998). A new intellectual framework for psychiatry? *American Journal of Psychiatry,* 155, p. 457.

34. Squire, L., Kandel, E. (1999). *Memory: From Mind to Molecules* (New York: Scientific American Press).

35. Rock, David & Schwartz, Jeffrey, (2006). The neuroscience of leadership, *Strategy + Business Magazine,* 43.

36. Luscher, C., et. al. (2000). Synaptic plasticity and dynamic modulation of the postsynaptic membrane. *Nature Neuroscience,* 3, p. 545.

37. Adapted from Luscher, C. et. al. (2000). Synaptic plasticity and dynamic modulation of the postsynaptic membrane. *Nature Neuroscience,* 3, p. 545.

38. Matus, A. (2000). Actin-based plasticity in dendritic spines. *Science,* 290, p. 754.

39. Ibid, Kandel. p. 457.

40. Vogel, G. (2000). New brain cells prompt new theory of depression. *Science,* 290, p. 258.

41. Sheline, Y., et. al. (1996). Hippocampal atrophy in recurrent major depression. *Proceedings of the National Academy of Sciences,* 93, p. 398.

42. Lisman, J., Morris, G. (2001). Why is the cortex a slow learner? *Nature,* 411, p. 248.

43. Seligman, Martin (2004). *Authentic Happiness* (New York: Free Press).

44. Ibid, Rossi, p. 125.

45. Bittman (2001), quoted in Rossi, p. 244.

46. Ibid, Rossi, p. 126.

47. Shors, T., et. al. (2001). Neurogenesis in the adult is involved in the formation of trace memories. *Nature,* 410, p. 372.

48. Waelti, P., et. al. (2001). Dopamine responses comply with basic assumptions of formal learning theory. *Nature,* 412, p. 43.

49. Ibid, Rossi, p. 140.

50. Whitehouse, David. (2003, July 1). Fake alcohol can make you tipsy. Summary of article in *Psychological Science,* published by the American Psychological Association, reported in *BBC News.*

51. Motluk, Alison (2005, August). Placebos trigger an opioid hit in the brain. *New Scientist* 22:00, p. 23.

52. *Science* (2003, November). Brain maps perceptions, not reality, p. 4.

53. Ibid, *Science.*

54. Davidson, Richard J., Kabat-Zinn, John, et. al. (2003). Alterations in immune function produced by mindfulness meditation. *Psychosomatic Medicine,* 65, p. 564.

55. Orme-Johnson, D. W., et. al. (2006, August). Neuroimaging of meditation's effect on brain reactivity to pain. *Neuroreport,* 17 (12), p. 1359.

56. Beck, Robert (1986). Mood modification with ELF magnetic fields: A preliminary exploration, *Archaeus*, p. 4.

57. Schumann, W. O. (1957), Elektrische Eigenschwingungen des Hohlraumes Erde-Luft-Ionosphare, *Z. angew. u. Phys*, 9, p. 373.

58. www.lifetechnology.org/schumann.htm, accessed Feb 3, 2007.

59. Roffey, Leane E. (2006, April). The bioelectric basis for "healing energies." *Journal of Non-Locality and Remote Mental Interactions*, Vol. 4, p. 1.

60. Shealy, C. N., and Church, D. (2006), *Soul Medicine* (Santa Rosa: Elite), p. 87.

61. Ibid, Roffey, p. 1.

62. Ibid, Rossi, p. 190.

63. *Academic Medicine* (1991), 99 (9), p. S4.

CH. 4 THE BODY PIEZOELECTRIC

1. itp.nyu.edu/-nql3186/electricity/pages/pliny.html

2. Fukada, E., & Yasuda, I. (1956). On the piezoelectric effect of bone. *Journal of the Physical Society of Japan*, 12, p. 1158.

3. Ibid, Shealy and Church, p 187.

4. Oschman, James (2003). *Energy Medicine in Therapeutics and Human Performance* (Edinburgh: Butterworth Heineman), p. 5.

5. Szent-Gyorgi, A. (1957). *Bioenergetics* (New York: Academic Press), p. 139, quoted in Oschman, p. 88.

6. Biscar, J.P. (1976). Photon enzyme action, *Bulletin of Mathematical Biology*, 38, p. 29.

7. Coetzee, H. (2000). Biomagnetism and bio-electromagnetism: the foundation of life. *Future History*, 8.

8. Kirschvink, J.L., et. al. (1992) Magnetite biomineralization in the human brain. *Proceedings of the National Academy of Sciences*, 89, p. 7683.

9. Ibid. Shealy & Church, p 191.

10. Grad, B. (1965). Some biological effects of "laying on of hands" *Journal of the American Society for Psychical Research*, 59, p. 95.

11. Ibid, Grad, pp. 95-127.

12. Grad, B., et. al. (1964). The influence of an unorthodox method of treatment on wound healing in mice. *International Journal of Parapsychology*, 3, p. 5.

13. Otto, H. A., Knight, J. W. (1979) *Dimensions in Wholistic Healing* (Chicago: Nelson-Hall) p. 199.

14. Maret, K. (2005). Seven key challenges facing science. *Bridges*, Spring, (16) 1, p. 14.

15. Sandyk, R., Anninos P.A., Tsagas, N., Derpapas, K. (1992), Magnetic fields in the treatment of Parkinson's disease. *International Journal of Neuroscience*, 63 (1-2) p. 141.

16. Liboff, A. R. (2004). Toward an electromagnetic paradigm for biology and medicine. *Journal of Alternative and Complimentary Medicine*, 10 (1), p. 41.

17. University of Illinois, Chicago (2006). The paradox of increasing genomic deregulation associated with increasing malignancy. www.mechanogenomics.com. *Accessed 8-1-06*

18. *The Economist* (2005, December 24). The human epigenome: life story, the sequel, p. 104.

19. *United Press International* (2006, January 31). Magnetic stimulator tested for depression.

20. Spence, Graham (2000) quoted in Eden, Donna, *Energy Medicine* (New York: Bantam) p. 299.

21. Childre, L, and Martin, H, (2000), *The HeartMath Solution*, (San Francisco: HarperSanFrancisco), p. 34.

22. Kelly, Robin (2000), *Healing Ways* (New Zealand: Penguin), p 153.

23. Shealy, C. N., and Church, D. (2006), *Soul Medicine* (Santa Rosa: Elite), p 190.

24. Ibid. Shealy & Church, p 190.

25. Ibid. Shealy & Church, p 194.

26. Shealy, C.N. (2005). *Life Beyond 100*. (New York: Tarcher), p. 152.

27. Schlebush, K.P., Walburg, M.O. Walburg, & Popp, F.A. (2005). Biophotonics in the infrared spectral range reveal acupuncture meridian structure of the body, *The Journal of Alternative and Complementary Medicine*, 11 (1), p. 171.

28. McTaggart, Lynne (2002). *The Field*. (New York: Quill), p. 55.

29. Hyvarien, J., and Karlssohn, M. (1977). Low-resistance skin points that may coincide with acupuncture loci, *Medical Biology*, Vol. 55, pp. 88-94, as quoted in *New England Journal of Medicine* (1995), Vol. 333, No. 4, p. 263.

30. Kober, A., et. al. (2002). Pre-hospital analgesia with acupressure in victims of minor trauma: a prospective randomized, double-blinded trial. *Anesthesia & Analgesia*, 95, p. 723.

31. Allen, John, et. al. (1998). The efficacy of acupuncture in the treatment of major depression in women. *Psychological Science,* 9(5), p. 397.

32. Molsberger, A.F., et. al. (2006, April 12). Designing an acupuncture study: the nationwide, randomized, controlled, German acupuncture trials on migraine and tension-type headache. *Journal of Alternative and Complementary Medicine,* 3, p. 237.

33. Rosenfeld, Isadore (2006). Does acupuncture really work? *Parade,* July 9, p. 17.

34. Cho, Z.H., et. al. (2006). Neural substrates, experimental evidences and functional hypthesis of acupuncture mechanisms. *Acta Neurobiological Scandinavica,* 113, p. 370.

35. Korotkov, K., Williams, B., Wisneski, L.A. (2004). Assessing biophysical energy transfer mechanisms in living systems. *Journal of Alternative and Complementary Medicine,* 10(1), p. 49.

36. *Alternative Medicine* (2002, March). High technology meets ancient medicine, p. 93.

37. ACEP (2006). *ACEP Certification Program Manual.*

38. Matthews, Ronald (2006). Harold Burr's biofields: Measuring the electromagnetics of life. *Paper submitted for publication.*

39. Burr, H.S. (1972). *The Fields of Life.* (New York: Ballentine).

40. Suarez, Edward (2004). C-reactive protein is associated with psychological risk factors of cardiovascular disease in apparently healthy adults. *Psychosomatic Medicine,* 66, p. 684.

41. Childre & Martin, (2000) *The HeartMath Solution,* (San Francisco: HarperSanFrancisco), p. 15.

42. Burr, H.S. (1957). Harold Saxton Burr, *Yale Journal of Biology and Medicine,* 30, p. 161.

43. Liu, Hong, et. al. (2002). Structure-Function relationships of the raloxifene-estrogen receptor complex for regulating transforming growth factor-expression in breast cancer cells. *Journal of Biological Chemistry,* 277(11), p. 9189.

44. Tiller, William (1997). *Science and Human Transformation* (California: Pavior), p. 120.

45. Gardner, S.E., et. al. (1999, November). Effect of electrical stimulation on chronic wound healing: a meta-analysis. *Wound Repair and Regeneration,* 7, p. 495.

46. Beneviste, J. et. al. (1992). Highly dilute antigen increases coronary flow of isolated heart from immunized guinea-pigs, *FASEB Journal,* 6, p. A 1610.

47. Beneviste, J., et. al. (1997). Transatlantic transfer of digitized antigen signal by telephone link. *Journal of Allergy and Clinical Immunology,* 99, p. S175.

48. Swanson, Claude (2003). *The Synchronized Universe* (Colorado Springs: Poseida).

49. Ballegaard, S., et al: Acupuncture in severe stable angina pectoris—a randomized trial. *Acta. Med Scand,* 220 (4), p. 307.

50. Shealy, C. N., Helms, J, McDaniels, A. (1990). Treament of male infertility with acupuncture. *The Journal of Neurological and Orthopaedic Medicine and Surgery, December,* 11 (4), p. 285.

51. Hanson, P.E., Hansen, J.H. (1985, September). Acupuncture treatment of chronic tension headache—a controlled cross-over trial. *Cephalgia,* 5 (3), p. 137.

52. Ibid. Shealy & Church, p. 214.

53. Feinstein, David (2007). Energy psychology: method, theory, evidence. Paper submitted for publication, p. 6. www.EnergyPsychologyResearch.com, accessed Feb 5.

54. Tiller, William (1997). *Science and Human Transformation* (California: Pavior), p. 119.

55. Yount, Garret, et. al. (2005). Changing perspectives on healing energy in traditional Chinese medicine. *Consciousness & Healing.* (St Louis: Elsevier), p. 424.

56. Wang C., Collet, J.P., & Lau, J. (2004, March 8). The effect of Tai Chi on health outcomes in patients with chronic conditions: a systematic review. *Archives of Internal Medicine,* 164 (5), p. 493.

57. Mayer, Michael (2004). *Secrets of Living Younger Longer* (San Francisco: Bodymind) p. 6.

58. Grauds, Constance (2005). *The Energy Prescription* (New York: Bantam), p. 94.

59. Ibid. *Alternative Medicine,* p. 93.

60. Rooke, A. (2006), Professor of Anesthesiology, University of Washington, *Pacemakers,* AVAA Forum.

61. Fenster, Julie M. (2003). *Mavericks, Miracles and Medicine* (New York: Barnes & Noble), p. 171.

62. *Neuroinsights* (2006), neurotechnology publication: www.neuroinsights.com/pages/24/index.htm

63. Shealy, C. N., Church, D. (2006). *Soul Medicine* (Santa Rosa: Elite).

64. Shealy, C. N. (2006). *Life Beyond 100.* (New York: Tarcher).

Ch. 5 The Connective Semiconducting Crystal

1. Adapted from Oschman, James (2003). *Energy Medicine in Therapeutics and Human Performance.* (Edinburgh: Butterworth Heineman), p. 93.

2. Ibid. Oschman p. 93.

3. Pickup, A. (1978). Collagen and behavior: A model for progressive debilitation. *Medical Science,* 6, p. 499, quoted in Oschman, p. 92.

4. Ho, Mae Wan (1999, October 2). Coherent energy, liquid crystallinity and acupuncture. Address to British Acupuncture Society. Quoted in Oschman, p. 87.

5. Ibid. Oschman, p. 92.

6. McCrone, John (2004, June 1), How do you persist when your molecules don't? *Science and Consciousness Review*. www.sci-con.org/articles/20040601.html

7. Chen, E. (2006, September). Are cytoskeletons the smallest electric power line to transmit electricity in the meridians? *Journal of Accord Integrative Medicine,* 2 (5), p. 109.

8. Ibid. Oschman, p. 213.

9. Ibid. Oschman, p. 213.

10. Lipton, Bruce (2005). Quoted by Ardagh, Arjuna, in *The Translucent Revolution* (Novato: New World Library), p. 341.

11. Savva, Savely (2006, September). Biofield control field of the organism—what is its physical carrier? *Journal of Accord Integrative Medicine,* 2 (5), p. 112.

12. Used with permission, adapted from Oschman p. 65.

13. Ibid. Oschman, p. 47.

14. Ibid. Oschman, p. 46.

15. McTaggart, Lynne. (2002). *The Field* (New York: Quill), p. 46.

16. Ibid. Oschman, p. 79.

17. Ahn, Andrew, et. al. (2005). Electrical impedance along connective tissue planes associated with acupuncture meridians. *Complementary and Alternative Medicine,* 5, p. 10.

18. Cho, Z. H. (1998) New findings of the correlation between acupoints and corresponding brain cortices using functional MRI, *Proceedings of National Academy of Science*, 95, p. 2670.

19. Adapted from Oschman, p. 79. Used with permission.

20. McClare, C.W.F. (1974). Resonance in bioenergetics, *Annals of the New York Academy of Science*, 227, p. 74.

21. Feinstein, David (2006). Six Pillars of Energy Medicine. *Paper submitted for publication.*

22. I am indebted to James Oschman for this insight at the 2006 ISSSEEM conference, Denver, CO, June, 24

23. McTaggart, Lynne (2002). *The Field.* (New York: Quill), p. 47.

24. Ibid, McTaggart, p. 48.

25. Frohlich, Herbert (1968). Long-range coherence and energy storage in biological systems. *International Journal of Quantum Chemistry,* 2, p. 641, quoted in McTaggart, p. 49.

26. Ibid, McTaggart, p. 49.

27. Ho, Mae Wan, quoted in Oschman, p. 310.

28. Ibid, Oschman, p. 72.

CH. 6 ANATOMY OF A MONSTOR

1. *The Economist* (2005, December 17). Scalpel, scissors, lawyer, p. 30.

2. Toner, Robin (2004, January 14). Democrats see a new urgency in health care. *New York Times.*

3. Anderson, G., et. al. (2005). Health spending in the United States and the rest of the industrialized world. *Health Affairs,* 24 (4), p. 903.

4. Ibid, Anderson, p. 903.

5. *The Economist* (2004, March 27). Crunch-time coming: Concern, but no action, on rising health-care costs for the elderly, p. 34.

6. Ibid, *The Economist*, p. 34.

7. *The Economist* (2006, January). Desperate measures: America's health-care crisis, p. 24.

8. Ibid, Anderson, p. 903.

9. Kerr, E.A., McGlynn, E.A., et al (2004). Profiling the quality of care in twelve communities: results from the CQI study. *Health Affairs,* 23 (3), p. 247.

10. *New York Times* (2005, March 6). One nation under stress, reprinted from *The Economist.*

11. *The Economist* (2005, December 17). Scalpel, scissors, lawyer, p. 30.

12. Weil, Andrew T., and Snyderman, Ralph. (2002). Integrative Medicine: Bringing medicine back to its roots, *Archives of Internal Medicine,* 162, p. 395.

13. Ibid, *The Economist*. p. 30.

14. *Sutter Your Health* (2005), Sutter Lakeside in the community. Summer, p. 4.

15. Murphy, Michael, and Donovan, Steven. (1988). *Contemporary Meditation Research:* A Review of Contemporary Meditation Research With a Comprehensive Bibliography, p. 1931. (San Francisco: The Esalen Institute).

16. Dossey, Larry (1997) summarizes many of them in *Prayer is Good Medicine* (San Francisco: Harper), p. 8.

17. McCraty, Rollin (2003). Impact of the power to change performance program on stress and health risks in correctional officers. *Institute of HeartMath Report* No. 03-014.

18. Reddy, B. S., Burrill, C. and Rigotty, J. (1991). Effect of diets high in omega-3 and omega-6 fatty acids on initiation and post-initiation stages of colon carcinogenesis. *Cancer Research*, 51, p. 487.

19. Simopoulos A. P. (1997). Omega-3 fatty acids in the prevention-management of cardiovascular disease. *Canadian Journal of Physiological Pharmacology,* 75 (3), p. 234. Also, Hu, Frank, B., et al. (2002). Fish and omega-3 fatty acid intake and risk of coronary heart disease in women. *Journal of the American Medical Association*, 287, p. 1815.

20. Lou, J., et al. (1996). Dietary (n-3) polyunsaturated fatty acids improve adipocyte insulin action and glucose metabolism in insulin-resistant rats: relation to membrane fatty acids. *Journal of Nutrition,* 126 (8), p.1951. Also, Borkman, M., et al. (1989) Effects of fish oil supplementation on glucose and lipid metabolism in NIDDM. *Diabetes*, 38 (10), p. 1314.

21. Ibid, Dozor, p. 309.

22. Ornish, Dean (2004). Love as healer: *The Heart of Healing* D. Church, Ed. (Santa Rosa: Elite Books), p. 258.

23. Houston, Jean (2000). *Jump Time: Shaping Your Future in a Time of Radical Change.* (New York: Tarcher/Putnam).

24. Ibid, Ornish, p. 258.

25. Oxman, Thomas E., et al. (1995). Lack of social participation or religious strength and comfort as risk factors for death after cardiac surgery in the elderly. *Psychosomatic Medicine*, Vol. 57, pp. 5-15.

26. Dossey, Larry (2005). Non-local consciousness and the revolution in medicine in *Healing our Planet, Healing Our Selves.* D. Church, Ed.(Santa Rosa: Elite Books), p 149.

Ch 7 Consiousness as Medicine

1. Ibid, Oxman, 5.

2. Davidson, Richard J., Kabat-Zinn, John, et al. (2003). Alterations in immune function produced by mindfulness meditation. *Psychosomatic Medicine,* 65, p. 564.

3. Seskevich, J.E., Crater, S.W., Lane, J.D., Krucoff, M.W. (2004). Beneficial effects of noetic therapies on mood before percutaneous intervention for unstable coronary syndromes. *Nursing Research,* 53 (2), p. 116.

4. Grunberg, G. E., et al. (2003). Correlations between preprocedure mood and clinical outcome in patients undergoing coronary angioplasty. *Cardiology Review* 11(6) p. 306.

5. Lipton, Bruce (2005). *The Biology of Belief* (Santa Rosa: Elite Books), p. 119.

6. Olshansky, J., quoted in Social security's latest obstacle: life expectancy. *New York Times,* Robert Pear byline, Dec 31, 2004.

Ch. 8 Belief Therapy

1. Ibid, Lipton, pp. 75-94.

2. Pierce, Joseph Chilton (2005). *The Biology of Belief* (Santa Rosa: Elite Books), Back cover comment.

3. Lipton, Bruce, et al. (1991). Microvessel endothelial cell transdifferentiation: phenotypic characterization. *Differentiation,* 46, p. 117.

4. Lipton, Bruce, et al. (1992). Histamine-modulated transdifferentiation of dermal microvascular endothelial cells. *Experimental Cell Research,* 199, p. 279.

5. *Science* (2001, August 10), Epigenetics special issue, 293:5532.

6. Lipton, Bruce (2005). *The Biology of Belief,* p. 112.

7. Oschman, James L. (2005). Energy and the healing response. *Journal of Bodywork and Movement Therapies* vol.9, p. 11.

8. McCraty, Rollin, et al. (2003). Modulation of DNA conformation by heart-focused intention (Boulder Creek: Institute of HeartMath). *HeartMath Research Center Publication 03-008.*

9. Hedges, Fran (2006). Students' stress busters, Roehampton University, United Kingdom: www.roehampton.ac.uk/stressbusters/overview.asp

10. Institute of HearthMath (2003). *Emotional Energetics, Intuition and Epigenetics Research.* (Boulder Creek: Institute of HearthMath), p. 1.

11. Childre, L., & Martin, H (2000). *The Heartmath Solution* (San Francisco: HarperSanFrancisco), p. 37.

12. Ibid, Childre & Martin.

13. McCraty, R., et. al. (2003). Modulation of DNA conformation by heart-focused intention. (Boulder Creek: Institute of HearthMath), p. 4.

14. Bairoch, A., et. al. (2005). The universal protein resource. *Nucleic Acids Research,* 33, p. D154.

15. Moseley, J.B. (2002, July 11). A controlled trial of arthroscopic surgery for osteoarthritis of the knee. *New England Journal of Medicine,* 347, p. 81.

16. Baylor College of Medicine Press Release. www.eurekalert.org/pub_releases/ 2002-07/bcom-sfc070802.php *Accessed 8-1-06*

17. Ibid, Shealy & Church, p. 172.

18. Vedantam, Shankar (2002, May 7). Against depression, a sugar pill is hard to beat. *Washington Post,* p. A01

19. Stein, R. & Kaufman, M. (2006, January 1). *Washington Post.*

20. Ibid, Shealy & Church, p. 182.

21. Moore, T.J. (1999, October 17). No prescription for happiness. *Boston Globe,* also www.motherjones.com/news/feature/2003/11/ma_565_01.html pn.psychiatryonline.org/cgi/content/full/39/14/1

22. Kelleher, Susan, & Wilson, Duff (2005, June 26). The hidden big business behind your doctor's diagnosis. *Seattle Times.*

23. Shealy, C.N. (2005). *Life Beyond 100* (New York: Tarcher Putnam), p. 27.

24. Maugh, Thomas (2006, July 21). Drug mistakes hurt or kill 1.5 million each year in US. *Los Angeles Times.*

25. Motluk, Alison (2005, December 8). 'Safe' painkiller is leading cause of liver failure. *New Scientist,* p. 19.

26. Gander, Marie-Louise, et. al. (2005, January - February). Post-traumatic stress disorder: a new risk factor for coronary artery disease? *Psychosomatic Medicine,* 67:1.

CH. 9 ENTANGLED STRINGS

1. Church, D. (2004). Journey of a Pomo Indian medicine man, in *The Heart of Healing* (Santa Rosa: Elite), p. 31.

2. For one catalog of studies of distant healing, see psychiatrist Daniel Benor's database at www. WholisticHealingResearch.org. For another, see the Institute of Noetic Sciences web site, specifically www.noetic.org/research/dh/studies.html (accessed 3/1/2007).

3. Sicher, F., and Targ, E., et. al. (1998). A randomized double-blind study of the effect of distant healing in a population with advanced AIDS. *Western Journal of Medicine,* 168 (6), p. 356.

4. Abbott, N. C. (2000). Healing as a therapy for human disease: a systemic review. *Journal of Alternative and Complementary Medicine,* 6 (2), p. 159.

5. Benor, Daniel (1990). Survey of spiritual healing research *Complementary Medical Research,* 4 (1), p. 9.

6. Jonas, W.B. (2003). Science and spiritual healing. *Alternative Therapies in Health and Medicine,* 9 (2), p. 56.

7. Goswami, Amit (2000). *The Visionary Window.* (Wheaton: Quest), p. 15.

8. Goswami, Amit (2004). *The Quantum Doctor* (Charlottesville: Hampton Roads), p. 62.

9. Ibid, Goswami, p. 15.

10. Ibid, Goswami, p. 62.

11. McTaggart, Lynne (2007), The Intention Experiment (New York: Free Press), p. xx.

12. Peoc'h, Rene (2002, September 1). Psychokinesis experiments with human and animal subjects upon a robot moving at random. *Journal of Parapsychology.*

13. Ibid, Goswami, p. 63.

14. Bruce, Colin (2004). *Schrodinger's Rabbits* (Washington: Joseph Henry), p. 90.

15. Ibid, Goswami, p. 64.

16. From Nova PBS documentary: *The Elegant Universe,* www.pbs.org/wgbh/ nova/elegant/everything.html. *Accessed 8-1-06*

17. BBC2 (2002, February). *Horizon* program London.

18. Cline, David (2003, March). The search for dark matter. *Scientific American.*

19. Swanson, Claude (2003). *The Synchronized Universe* (Colorado Springs: Poseida).

20. Kaku, Michio (2006, June 24). Reading the mind of God. International Society for the Study of Subtle Energies and Energy Medicine (ISSSEEM). Denver, CO, conference address.

21. Radin, Dean (2006) *Entangled Minds* (New York: Pocket), p. 1.

22. Brooks, M (2004, March 27). The weirdest link. *New Scientist.*

23. Ibid, Radin, p. 17.

24. Boehm, D., & Hiley, B. (1992). *The Undivided Universe.* (London: Routledge), p. 382.

25. Kelly, Robin (2006). *The Human Aerial.* (New Zealand: Zenith), p. 89.

26. *The Economist* (2005, May 14). A mirror to the world, p. 81.

27. Ibid, *The Economist*, p. 81.

28. Einstein, A., Podolsky, B., Rosen, N., (1935) Can quantum mechanical description of reality be considered complete? *Physical Review,* 47, pp. 777- 780.

29. Grinberg-Zylberbaum, J., et. al. (1994), The Einstein-Podolsky-Rosen paradox in the brain: the transferred potential, *Physics Essays,* 7, p. 422.

30. Von Franz, Marie-Louse, quoted in McCallum, Ian (2005). *Ecological Intelligence,* (Cape Town: Africa Geographic), p. 146.

31. Quoted in Dossey, Larry (2001) *Healing Beyond the Body* (Boston: Shambhala), p. 199.

Ch. 10 Scanning the Future

1. Quoted in Dossey (1997). *Prayer Is Good Medicine* (New York: HarperCollins), p. 11.

2. Marianoff, Dmitri (1944). *Einstein: An Intimate Study of a Great Man* (New York: Doubleday).

3. McCraty, Rollin (2005, January). Telephone and email exchanges.

4. Radin, D. I., Taylor, R. D. & Braud, W. (1995). Remote mental influence of human electrodermal activity: A pilot replication. *European Journal of Parapsychology*, 11, p. 19.

5. McCraty, Rollin, et al. (2004). Electrophysiological evidence of intuition: Part 1. The surprising role of the heart. *Journal of Alternative and Complementary Medicine,* 10 (1), p. 133.

6. Durka, P. et. al. (2000). Time-frequency microstructure of event-related desynchronization and synchronization. *Medical and Biological Engineering and Computing,* 39, p. 315.

7. Kelly, Robin (2006). The Human Aerial (New Zealand: Zenith), p. 96.

8. McCraty, Rollin, et al. (2004). Electrophysiological evidence of intuition: Part 2: A system-wide process? *Journal of Alternative and Complementary Medicine,* 10 (2), p. 325.

9. Arden, John (1998). *Science, Theology and Consciousness* (Westport: Praeger), p. 104.

10. Norretranders, Tor (1998). *The User Illusion* (New York: Viking Penguin), p. 216.

11. Ibid, Norretranders, 259.

12. Blanton, Brad (2000). *Practicing Radical Honesty* (Stanley;VA Sparrowhawk).

13. Penrose, Roger (1990). *The Emperor's New Mind* (New York: Vintage), p. 574.

Ch. 11 Routine Miracles

1. Quoted in *Dr. Andrew Weil's Self-Healing* (2006, July), p. 2.

2. Dillard, James (2004). Pain as our greatest teacher. In *The Heart of Healing.* D. Church, Ed. (Santa Rosa: Elite Books), p. 247.

3. Dillard, James (2003). *The Chronic Pain Solution* (New York: Bantam).

4. Dillard, James (1998). *Alternative Medicine for Dummies* (Palo Alto: IDG).

5. Jewish Theological Seminary (2004). Survey of physicians' views on miracles. (New York: Jewish Theological Seminary). www.jtsa.edu/research/finkelstein/surveys/physicians.shtml. *Accessed 8-1-06*

6. Woodcock, Alexander, & Davis, Monte (1978). *Catastrophe Theory* (New York: Viking Penguin), 9, quoted in Miller, William, & C'deBaca, Janet (2001). *Quantum Change* (New York: Guilford), p. 71.

7. Miller, William, & C'deBaca, Janet (2001). *Quantum Change* (New York: Guilford).

8. Shealy, C. N. (1999). *Sacred Healing* (Boston: Element), p. xv.

9. Ibid, Greenfield, p. 186.

10. Maret, Karl (2005). Seven key challenges facing science. *Bridges,* Spring, p. 7.

Ch. 12 Meridian Based Therapies

1. Robins, Eric (2004). *The Heart of Healing* D. Church, Ed. (Elite, Santa Rosa), p. 281.

2. Oschman, James L. (2006) Trauma energetics. *Journal of Bodywork and Movement Therapies,* vol. 10, p. 32.

3. Wells, Steven, et al. (2003) Evaluation of a meridian-based intervention, emotional freedom techniques (EFT), for reducing specific phobias of small animals. *Journal of Clinical Psychology*, 59 (9), p. 943.

4. Wells, Steven (2005). *Email to author dated March 27.*

5. Baker, H., Siegel, L. (2005). *Can a 45 minute session of EFT lead to reduction of intense fear of rats, spiders and water bugs?* Reported at ACEP Convention, Baltimore, MD. Paper submitted for publication.

6. Rowe, Jack (2005, September). The effects of EFT on long-term psychological symptoms. *Counseling and Clinical Psychology Journal*, 2 (3), p. 104.

7. Temple, Graham (2006). *Reducing anxiety in dental patients with EFT.* www.emofree.com/Research/graham-temple-dental-study.htm.

8. Lambrou, P. T., et. al., (2003). Physiological and psychological effects of a mind/body therapy on claustrophobia. Subtle Energies and Energy Medicine, 14(3), p. 239.

9. Kober, A., et. al, (2002). Pre-hospital analgesia with acupressure in victims of minor trauma: A prospective, randomized, double-blind trial. Anesthesia & Analgesia, 95(3), p. 723.

10. A list of peer reviewed studies of EMDR appears on the University of Buffalo's web site at www.socialwork.buffalo.edu/fas/smyth/Personal_Web/ EMDR/EMDR_Internet_Articles.htm

11. Feinstein, David (2007). Energy psychology: method, theory, evidence. www.EnergyPsychologyResearch.com, accessed Feb 5.

12. Ibid, Feinstein, p. 297.

13. Ibid, Feinstein, back cover. Used with permission.

14. Oschman, James (2006). Trauma energetics. *Journal of Bodywork and Movement Therapies,* 10, p. 32

15. Ibid, Feinstein, p. 322.

16. Russek, L., & Schwartz, G. E. (1997). Perceptions of parental love and caring predict status in midlife: A 35-year followup of the Harvard mastery of stress study. *Psychosomatic Medicine,* 59 (2), p. 144.

17. Ibid, Feinstein, p. 109.

18. Oschman, James (2006). Trauma energetics. *Journal of Bodywork and Movement Therapies,* 10, p. 21

19. www.emofree.com/Research/rouleaux.htm *Accessed 8-1-06*

20. For a listing of studies see David Benor's list at www.emofree.com/cousins/ benor.htm and also www.childtrauma.com/pub.html

21. Feinstein, D., Craig, G., Eden, D. (2005). *The Promise of Energy Psychology* (New York: Tarcher).

22. Ibid, Feinstein, p. 67.

23. Ibid, Feinstein, p. 108.

24. Benor, Daniel (2006, February 15). Teleclass with author.

25. Benor, Daniel (2006). www.wholistichealingresearch.com/Articles/ Selfheal.asp. *Accessed 8-1-06*

26. Temes, Roberta (2006). *The Tapping Cure* (New York: Marlowe) p. 140.

27. Lipton, Bruce (2005). *The Biology of Belief* (Santa Rosa: Elite Books), p. 84.

28. Williams, Robert (2004). *Psych-K: the Missing Peace in Your Life.* (Colorado: Myrddin), p. 85.

29. Testimonials at www.psych-k.com/real_results.php

30. Fleming, Tapas (2006, May 5). Emotional healing with tapas acupressure technique. Presentation at 8th Annual International Energy Psychology Conference, Santa Clara, CA.

31. Ibid, Fleming, May 5.

32. Stoler, Larry (2006, May 30). Telephone conversation with author.

33. Barrett, Bruce (2006). Placebo, meaning, and health. *Perspectives in Biology and Medicine*, Spring, 49 (2), p. 178.

34. Ibid, Yount, p. 423.

CH. 13 SOUL MEDICINE AS CONVENTIONAL MEDICINE

1. Eccles, John (1991). *Evolution of the Brain, Creation of the Self* (New York: Routledge)

2. Ibid, Shealy & Church, p. 74.

3. University of Illinois, Chicago (2006). The paradox of increasing genomic deregulation associated with increasing malignancy. www.mechanogenomics.com.

4. Hubbard, Barbara Marx (2005). Bringing god home. In *Healing Our Planet, Healing Our Selves.* (Santa Rosa: Elite Books), p. 11.

5. Burnham, Terry, & Phelan, Jay (2000). *Mean Genes* (Cambridge: Perseus), p. 220.

6. Ibid, Shealy & Church, p. 43.
7. Connor, Steve (2003, December 8). Glaxo chief: our drugs do not work on most patients, in the *Independent*.
8. Kelleher, Susan, and Wilson, Duff (2005, June 26). The hidden big business behind your doctor's diagnosis. *Seattle Times*.
9. Ibid, Kelleher.
10. Vedantam, Shankar (2006, April 12). Comparison of schizophrenia drugs often favors firm funding study. *Washington Post*.
11. Grauds, Constance (2005). *The Energy Prescription* (New York: Bantam), p. 15.

Ch. 14 Medicine for the Body Politic

1. Pamela Ney-Noyes was the practitioner.
2. Feinstein, D., Craig, G., Eden, D. (2005). *The Promise of Energy Psychology* (New York: Tarcher), p. 111.
3. Centers for Disease Control: www.cdc.gov
4. Ibid, CDC.
5. health.dailynewscentral.com/content/view/0001907/49. *Accessed 8-1-06.*
6. Nicosia, Greg (2006, May 6). ACEP Humanitarian Project . Annual International Energy Psychology Conference, Santa Clara, CA.
7. Grille, Robin (2005). Parenting for a Peaceful World (New Zealand: Longueville).
8. Levitt, Steven D. (2006), Freakonomics: A Rogue Economist Explores the Hidden Side of Everything, revised edition. (New York: Morrow).
9. Raine, A., et. al. (1994). Birth complications combined with early maternal rejection at age 1 year predispose to violent crime at 18 years. Archives of General Psychiatry, Vol 51, p. 984.
10. Odent, Michel (2006). The long term consequences of how we are born. Journal of Prenatal & Perinatal Psychology & Health. Vol 21, no 2. p. 187.
11. *Knight Ridder Newspapers* (2006). U.S. says health care spending will double to $4 trillion a year over decade. Jan 9.
12. E-mail to Author Feb 3, 2007.
13. *The New York American* (1909, June 20).
14. Photo courtesy of Yannis Behrakis/Reuters.
15. Johnson, C., et. al. (2001). Thought field therapy—soothing the bad moments of Kosovo. *Journal of Clinical Psychology,* 57, p. 1237.
16. Ibid, Feinstein, p. 16.
17. Feinstein, D. (2006, May 6) Annual International Energy Psychology Conference. Santa Clara, CA.
18. Cool, L.C. (2004, May). The power of forgiving. *Reader's Digest*, p. 94.
19. www.emdrhap.org/aboutus/ourefforts/middleeast.php. *Accessed* 8-1-06.
20. Sakai, C., et. al. (2001). Thought field therapy clinical application: Utilization in an HMO in behavioral medicine and behavioral health services. *Journal of Clinical Psychology,* 57, p. 1215.
21. Feinstein, David (2006), www.energypsych.org/studies.
22. Cane, Patricia (2000). *Trauma Healing and Transformation* (Santa Cruz: Capacitar), p. 10.
23. Ibid, Cane, p. 61.
24. Houston, Jean (2000). *Jump Time: Shaping Your Future in a Time of Radical Change* (New York: Tarcher/Putnam).

Ch. 15 Ten Principles of Epigenetic Medicine

1. Schakut, Anne (2005). The biomarker benefit. *Economist* Intelligent Life report, Summer, p. 106.
2. The Economist (2005, Sept 17). Medicine without frontiers. *Economist Technology Quarterly,* p. 37.
3. Alsever, Jennifer (2006, October). The gene screen, *Busness 2.0,* p. 110.
4. United Press International, (2007). New nanotechnology is announced, in *Science Daily.* Jan 24.
5. Fischbach, Amy (2005). YMCA customizes program to prolong exercise habits Associated Press, Aug 1, quoting a US study published in the *European Journal of Sports Science* (2003).
6. Collinge, William, in Kelly Robin (2000). Healing Ways (New Zealand: Penguin), p. 159.

7. Weil, Andrew T., and Snyderman, Ralph (2002). Integrative Medicine: Bringing medicine back to its roots, *Archives of Internal Medicine,* 162, p. 397.
8. Ibid. Weil, p. 397.
9. Weil, Andrew (1995). *Spontaneous Healing* (New York: Knopf), p. 226.
10. Ibid, Weil, p. 226.
11. Fugh-Berman, Adriane, et. al. (2002). *The Truth About Hormone Replacement Therapy* (New York: Prima).
12. Shealy, C. N. (2005). *Life Beyond 100* (New York: Tarcher), p. 37.
13. Shealy, C. N. (1999). *Sacred Healing* (Boston: Element), p. 102.
14. Siegel, Bernard (1988). *Love, Medicine and Miracles* (New York: Bantam).
15. www.teleosis.org/

CH. 16 PRACTICES OF EPIGENETIC MEDICINE

1. *Science* (2001, August 10), *Epigenetics* special issue, 293, p. 5532.
2. Ibid, Shealy, p. 36.
3. American College of Emergency Physicians (2006). *Seconds Save Lives* (Washington: ACEP).
4. Dossey, Larry (2006). *The Extraordinary Healing Power of Ordinary Things* (New York: Harmony)
5. Johansson, J. E., et. al. (1992, April 22). High 10-year survival rate in patients with early, untreated prostatic cancer. *Journal of the American Medical Association,* 267 (16).
6. Sukovich, Willliam (2006). New surgical trends in treating low back pain. *Martha Jefferson Hospital Clinical Front,* Spring, p. 1.
7. Weil, Andrew (2006, July). *Dr. Andrew Weil's Self-Healing,* p. 4.
8. Ibid, Weil, p. 6.
9. Swartz. Susan (2005, January 18). Managing menopause, in the *Santa Rosa Press Democrat,* p. D1.
10. Lewis, T., et. al. (2000) *A General Theory of Love* (New York: Random House), p. 221.
11. Ibid, Lewis, p. 222.
12. Barrett, Bruce (2006). Placebo, meaning, and health. *Perspectives in Biology and Medicine,* Spring, 49 (2), p. 178.
13. Channick, Steven (2006, January 18). Have physicians given up bedside manner in the quest for money? *HCD Health.*
14. Ibid, Shealy, 44.
15. Micozzi, Marc (2005, March 23). *Frontiers of Healing Forum.* Institute of Noetic Sciences.
16. Krames, Elliot. Pain Medicine—Using the Tools of the Trade, Pacific Pain Treatment Centers, San Francisco, www.painconnection.org/ MyTreatment/support_krames_0403.asp. *Accessed 8-1-06*
17. Ibid, Krames.
18. Blanton, Brad (1996). *Radical Honesty* (New York: Dell), p. 241.
19. Pearsall, Paul (2002). *The Beethoven Factor* (Charlottesville: Hampton Roads).
20. Pearsall, Paul (2004). The Beethoven factor. In *The Heart of Healing* D. Church, Ed. (Santa Rosa: Elite Books), p. 125.
21. Bahan, William (1988). *Spirit of Sunrise* (Denver: Foundation House).
22. Bruce, Eve (2002). *Shaman, M.D.: A Plastic Surgeon's Remarkable Journey into the World of Shapeshifting* (Boston: Inner Traditions).
23. Gould, Stephen Jay (1994, October). The evolution of life on earth. *Scientific American.*
24. Crick, Francis (1994). *The Astonishing Hypothesis: The Scientific Search for Soul* (New York: Scribner's), p. 3.

APPENDICES

1. Sha, Z. (2006) in *Healing the Heart of the World,* D. Church, Ed (Santa Rosa: Elite), p. 109.
2. Benor, Daniel (1994). www.WholisticHealingResearch.com/Articles/intuitiveassessmentsoverview.html *Accessed 8-1-06*
3. Shealy, C. N. (1996). *Ninety Days to Stress-Free Living* (London: Vega).
4. God, O.M.S. (±2005 BCE). *Holy Bible,* 2 Kings 5: v. 10-14.

Index